2006 IRC®

Q&A

APPLICATION GUIDE CHAPTERS 1-10

2006 IRC Q&A
Application Guide

Publication Date: March 2008
First Printing

ISBN 978-1-58001-666-9

Project Editor:	Greg Dickson
Layout Design:	Yolanda Nickoley
Cover Design:	Linda Cohoat
Typesetter:	Yolanda Nickoley
Production Coordinator:	Mary Lou Luif

Printed in the United States of America

Preface

The *2006 International Residential Code Q & A Application Guide* provides answers to commonly asked questions that arise in the application of Chapters 1 through 10 of the *2006 International Residential Code*® (IRC®). In addition, the guide includes photographs and illustrations to help provide an understanding of the intent of the code text. While not addressing every code section, this publication does attempt to address the most common requests for opinions regarding code issues. Also included are code topics that are of special importance or have gone through major revisions from previous code editions.

The *2006 International Residential Code Q & A Application Guide* is an essential resource for anyone involved with the IRC building provisions, including building officials, inspectors, plans examiners, architects, engineers, designers, builders, instructors, and students. The guide addresses those code provisions frequently encountered during design, plan review, construction, and daily code enforcement of the IRC.

The questions and answers found in this publication represent inquiries about code issues gathered over a period of years. Subjects are grouped by chapter in the same order as in the IRC. Each question is preceded by the language in the code section and followed by the answer. Many code sections contain a series of related questions and answers.

The *2006 International Residential Code Q & A Application Guide* is best used as a companion to the IRC. The code should always be referenced in order to gain a more comprehensive understanding of the IRC and its application.

Peter Kulczyk, International Code Council® Technical Staff, developed and compiled the material for this publication. He has over 23 years of experience in the application and enforcement of building codes and nine years of construction experience.

John Henry, P.E., ICC® Principal Staff Engineer, and Steve Van Note, ICC Senior Technical Staff, contributed with thorough reviews of the text.

The applications and illustrations published herein represent opinions of the authors and are not binding on the authority having jurisdiction. The authority having jurisdiction has the ultimate responsibility for rendering interpretations of the code.

TABLE OF CONTENTS

ADMINISTRATION

CHAPTER 1

R101 TITLE, SCOPE AND PURPOSE

R101.1 Title. These provisions shall be known as the *Residential Code for One- and Two-family Dwellings* of [NAME OF JURISDICTION], and shall be cited as such and will be referred to herein as "this code."

R101.2 Scope. The provisions of the *International Residential Code for One- and Two-family Dwellings* shall apply to the construction, alteration, movement, enlargement, replacement, repair, equipment, use and occupancy, location, removal and demolition of detached one- and two-family dwellings and townhouses not more than three stories above-grade in height with a separate means of egress and their accessory structures.

Q: A builder has submitted a plan for a six-unit back-to-back multiple dwelling project, and he has labeled it as a townhouse. We believe that it does not meet the complete definition of "Townhouse," which requires "two sides open," and thus we feel that it needs to be designed, permitted and constructed using the provisions of the *International Building Code*® (IBC®). Are we correct in our assumption?

A: Yes. it does not meet the scope of Section R101.2, as the two interior units are only open on one side.

Q: Does ownership affect the code requirements for dwellings or townhouses?

A: No. Many states have statutes related to ownership rights of occupants of multiple dwelling unit structures, but these statutes will have no affect on the code requirements in the IRC. The presence or lack of a true lot line can make a difference based upon the fire separation distance provisions.

Q: Does the IRC require a design professional, such as a licensed architect, for the design of a dwelling or townhouse meeting the scope of the IRC?

A: No. However, most states have statutes that regulate when a design professional is required, which may include some buildings designed under the IRC.

Q: The IBC requires an automatic sprinkler system in buildings housing residential occupancies. Does this mean that dwellings and townhouses meeting the scope of the IRC also require an automatic sprinkler system?

A: No. There are no provisions in the IRC that would require an automatic sprinkler system throughout. If a structure was outside the scope of the IRC, such as a proposed multiple-family structure that did not meet the definition of "Townhouse" in Section R202, then the building would need to be designed and constructed using the provisions in the IBC. In this case, a sprinkler system is required.

Q: The IBC classifies a residential dwelling as a Group R-3 occupancy, and an apartment building as a Group R-2 occupancy. The scope of the IRC makes reference to dwellings and townhouses, but it does not appear to indicate an occupancy classification. Where in the IRC is this information provided?

A: The IRC does not classify the occupancy with a letter/number designation as in the IBC.

SECTION R102 APPLICABILITY

R102.1 General. Where, in any specific case, different sections of this code specify different materials, methods of construction or other requirements, the most restrictive shall govern. Where there is a conflict between a general requirement and a specific requirement, the specific requirement shall be applicable.

R102.2 Other laws. The provisions of this code shall not be deemed to nullify any provisions of local, state or federal law.

R102.3 Application of references. References to chapter or section numbers, or to provisions not specifically identified by number, shall be construed to refer to such chapter, section or provision of this code.

R102.4 Referenced codes and standards. The codes and standards referenced in this code shall be considered part of the requirements of this code to the prescribed extent of each such reference. Where differences occur between provisions of this code and referenced codes and standards, the provisions of this code shall apply.

> **Exception:** Where enforcement of a code provision would violate the conditions of the listing of the equipment or appliance, the conditions of the listing and manufacturer's instructions shall apply.

Q: In Section R102.4, "Referenced codes and standards," it refers to standards that have been incorporated into the code. What is the purpose of these referenced standards?

A: The purpose of a standard is to establish nationally recognized requirements for the manufacture of a product and to provide the public with an understanding of the characteristics of that product. In the early 1920s the U.S. Department of Commerce instituted a plan to allow the industry to somewhat self-regulate itself by participating in the development of standards for the particular industry. The basis of this idea was that the manufacturers of a particular product would be the most knowledgeable about that product. So, during the balance of the last century, standards have been written for both the manufacturing process and installation of wood structural panels, chemically preservative-treated wood, roof shingles, mixing of mortar (including cold-weather provisions), vinyl siding, felt papers and house sheathing papers, expanded metal lath of exterior plaster and thousands of other products. These standards are being updated on a regular basis by the industry. The referenced standards in the code are contained in Chapter 43.

R102.5 Appendices. Provisions in the appendices shall not apply unless specifically referenced in the adopting ordinance.

SECTION AK101
GENERAL

AK101.1 General. Wall and floor-ceiling assemblies separating dwelling units including those separating adjacent townhouse units shall provide air-borne sound insulation for walls, and both air-borne and impact sound insulation for floor-ceiling assemblies.

Q: A townhouse project is proposed in our municipality. Fire separation walls are provided between the dwelling units as required by Section R317. Does the IRC have any requirement related to sound transmission between these dwelling units?

A: The provisions for sound transmission are contained in Section AK101, of Appendix K, "Sound Transmission." The appendices do not apply unless specifically referenced in the adopting ordinance of the municipality.

SECTION R103
DEPARTMENT OF BUILDING SAFETY

R103.1 Creation of enforcement agency. The department of building safety is hereby created and the official in charge thereof shall be known as the building official.

R103.2 Appointment. The building official shall be appointed by the chief appointing authority of the jurisdiction.

R103.3 Deputies. In accordance with the prescribed procedures of this jurisdiction and with the concurrence of the appointing authority, the building official shall have the authority to appoint a deputy building official, the related technical officers, inspectors, plan examiners and other employees. Such employees shall have powers as delegated by the building official.

Q: What are the minimum requirements for education and certifications for building inspectors working under the appointed building official?

A: There are no minimum requirements in the code; however, experience, education and certification are all factors to be considered. The building official is empowered to appoint other staff members, such as building inspectors, plans examiners and other employees to carry out the functions of the department based upon criteria established by the jurisdiction and the appointing authority.

SECTION R104 DUTIES AND POWERS OF THE BUILDING OFFICIAL

R104.1 General. The building official is hereby authorized and directed to enforce the provisions of this code. The building official shall have the authority to render interpretations of this code and to adopt policies and procedures in order to clarify the application of its provisions. Such interpretations, policies and procedures shall be in conformance with the intent and purpose of this code. Such policies and procedures shall not have the effect of waiving requirements specifically provided for in this code.

Q: The IRC directs the building official to enforce the provisions of the code. Can a local municipality waive requirements in the IRC, such as those required for the fire resistance of the exterior wall related to the location of a dwelling?

A: Under the provisions of the model code (IRC), no, there is no authority to waive requirements. However, the local or state governmental jurisdiction may amend the model code to fit its needs based on any number of factors, and these amendments may include specific provisions empowering the jurisdiction to waive requirements. This authority is only achieved through the adopting legislation in accordance with local or state law. The municipality also has the option to adopt any or all of the appendices.

R104.11 Alternative materials, design and methods of construction and equipment. The provisions of this code are not intended to prevent the installation of any material or to prohibit any design or method of construction not specifically prescribed by this code, provided that any such alternative has been approved. An alternative material, design or method of construction shall be approved where the building official finds that the proposed design is satisfactory and complies with the intent of the provisions of this code, and that the material, method or work offered is, for the purpose intended, at least the equivalent of that prescribed in this code. Compliance with the specific performance-based provisions of the International Codes in lieu of specific requirements of this code shall also be permitted as an alternate.

Q: In our municipality, there are many decks being constructed with composite materials. The code does not contain any definition that fits the criteria of these decks, and the prescriptive span tables in the code don't include composite materials. On what basis can these products be approved by our building inspection department?

A: These composite decking materials can be considered for approval by the building official under the "alternate materials" section of the code (Section R104.11). When using these products, the designer/builder or building inspector should evaluate the product for issues related to wood preservation, exposure to the ultraviolet rays of the sun, expansion and contraction during the temperature extremes, resistance to termites, surface-burning characteristics, flexural properties, type/size/spacing of fasteners, span ratings and other necessary properties. The ICC Evaluation Service (ICC ES) has established the *Acceptance Criteria for Thermoplastic Composite Lumber Products*, AC 109, that many of the manufacturers of these products have obtained. ICC ES reports that are issued based on AC 109 will list the scope of the report, properties evaluated, uses, description of the material, design and installation and conditions of use—all the information that will assist the building official in making the decision on whether to approve the product.

SECTION R105 PERMITS

R105.1 Required. Any owner or authorized agent who intends to construct, enlarge, alter, repair, move, demolish or change the occupancy of a building or structure, or to erect, install, enlarge, alter, repair, remove, convert or replace any electrical, gas, mechanical or plumbing system, the installation of which is regulated by this code, or to cause any such work to be done, shall first make application to the building official and obtain the required permit.

R105.2 Work exempt from permit. Permits shall not be required for the following. Exemption from permit requirements of this code shall not be deemed to grant authorization for any work to be done in any manner in violation of the provisions of this code or any other laws or ordinances of this jurisdiction.

Building:

1. One-story detached accessory structures used as tool and storage sheds, playhouses and similar uses, provided the floor area does not exceed 120 square feet ($11.15\ m^2$).
2. Fences not over 6 feet (1829 mm) high.
3. Retaining walls that are not over 4 feet (1219 mm) in height measured from the bottom of the footing to the top of the wall, unless supporting a surcharge.
4. Water tanks supported directly upon grade if the capacity does not exceed 5,000 gallons (18 927 L) and the ratio of height to diameter or width does not exceed 2 to 1.

5. Sidewalks and driveways.
6. Painting, papering, tiling, carpeting, cabinets, counter tops and similar finish work.
7. Prefabricated swimming pools that are less than 24 inches (610 mm) deep.
8. Swings and other playground equipment.
9. Window awnings supported by an exterior wall which do not project more than 54 inches (1372 mm) from the exterior wall and do not require additional support.

Q: There is a project in our municipality that consists of a series of existing older detached duplex dwellings forming a large residential complex. Alterations and renovations are proposed that involve foundation repair, new foundation waterproofing, siding repairs, reroofing and replacement of some structural members of the attached decks. Are building permits required?

A: Yes. Section R105.1 requires that all construction work must be controlled and regulated through the building code process. The only work that is exempted from the permit process are those specific items listed in Section R105.2. None of the proposed work you have cited is exempt. The code contains prescriptive requirements and provisions for this work in the form of material descriptions and properties, installation and fastening details, structural tables and performance-based goals (such as meeting loading requirements for wind, live and dead loads). Ordinary repairs are not applicable to this section. Ordinary repairs would involve simple repairs to limited areas and portions of the structures that are deteriorated or worn out. The building official may need to evaluate the particular situation to determine what permits are required.

R105.3.1 Action on application. The building official shall examine or cause to be examined applications for permits and amendments thereto within a reasonable time after filing. If the application or the construction documents do not conform to the requirements of pertinent laws, the building official shall reject such application in writing, stating the reasons therefor. If the building official is satisfied that the proposed work conforms to the requirements of this code and laws and ordinances applicable thereto, the building official shall issue a permit therefor as soon as practicable.

Q: In Section R105.3.1, it states that the building official shall examine the permit application "within a reasonable time after filing." What is a reasonable time?

A: Although the code does not specify what a reasonable time is, every municipality should consider what is reasonable in its area. The permit fee and plan review fee are intended to be used to staff those areas of the building department. For some municipalities, they may decide that single family dwellings should be able to go through the entire review process and permit issuance in two weeks, and for others it may be shorter or longer.

SECTION R109
INSPECTIONS

R109.1 Types of inspections. For onsite construction, from time to time the building official, upon notification from the permit holder or his agent, shall make or cause to be made any necessary inspections and shall either approve that portion of the construction as completed or shall notify the permit holder or his or her agent wherein the same fails to comply with this code.

R109.1.1 Foundation inspection. Inspection of the foundation shall be made after poles or piers are set or trenches or basement areas are excavated and any required forms erected and any required reinforcing steel is in place and supported prior to the placing of concrete. The foundation inspection shall include excavations for thickened slabs intended for the support of bearing walls, partitions, structural supports, or equipment and special requirements for wood foundations.

Q: Does the code contain any requirements for a timely inspection after the inspection is requested?

A: No. Although the code does not specify a time period for inspections, one could assume that there is an expectation by the permit holder that the inspector will show up at the scheduled time. Many municipalities establish written policies for inspection procedures that would provide the permit holder with an idea of what to expect. The administrative provisions in the code also require inspections at certain intervals during the construction process before the work can proceed.

SECTION R110 CERTIFICATE OF OCCUPANCY

R110.1 Use and occupancy. No building or structure shall be used or occupied, and no change in the existing occupancy classification of a building or structure or portion thereof shall be made until the building official has issued a certificate of occupancy therefor as provided herein. Issuance of a certificate of occupancy shall not be construed as an approval of a violation of the provisions of this code or of other ordinances of the jurisdiction. Certificates presuming to give authority to violate or cancel the provisions of this code or other ordinances of the jurisdiction shall not be valid.

Q: In our municipality the issuance of a certificate of occupancy is the responsibility of the zoning administrator, not the building official. Is this common practice in other parts of the country?

A: The IRC assigns authority and responsibility for administration and enforcement of the code to the building official, a defined term in the IRC representing the officer in charge. Included in the duties of the building official is the issuance of certificates of occupancy. This is an appointed position in the government jurisdiction. It is certainly desirable that the officer in charge have the experience, knowledge and qualifications necessary for this complex and important role in protecting public safety (Appendix A of the IBC offers model qualifications for a building official for use as a guide). In practice, jurisdiction job titles and duties vary greatly and do not necessarily match those in the model code. The officer in charge may carry the title of Chief Building Inspector, Director of Building Inspections, Building & Zoning Administrator and any number of other titles. Particularly in smaller jurisdictions, the person in charge of the building code may have many other responsibilities as well. In answer to your question about common practice, it varies greatly. Whether through the adopting legislation or by appointment, it is the government jurisdiction that decides who to place in the role of administering the code and issuing certificates of occupancy.

Q: Are interior finishes, such as carpet, tile, and wall coverings (paint or wallpaper) required to be installed for the dwelling final inspection?

A: No. There are no provisions in the code that require carpet, tile, paint or wallpaper to be installed.

Q: Are there any provisions in the code that require the installation of interior gypsum board prior to the dwelling's final inspection?

A: Yes, when the gypsum board is used to comply with fire-resistance requirements. An attached garage needs to be separated from the dwelling by ½-inch gypsum board per Section R309.1.12. Enclosed accessible space under stairs requires the application of ½-inch gypsum board per Section R311.2.2. Foam plastic insulation requires the application of ½-inch gypsum board or equivalent per Section R314. Fire-resistant separations must be complete for all townhouses and two-family dwellings per Section R317. Exterior walls also must meet the fire separation distance requirements of Section R302.

Q: Can a municipality require a garage to have a concrete floor?

A: Section R309.3 only states that garage floor surfaces be of approved noncombustible material and the material facilitate the movement of liquids toward the drain or vehicle entry door. The building official is the authority who can approve such material. The building official could determine that concrete is the only material he or she will approve to comply with these provisions, and he or she could require these as a condition of the certificate of occupancy.

Q: Can a temporary certificate of occupancy be issued by the building official?

A: Section R109.1.6 requires that a final inspection be made after the permitted work is complete and prior to occupancy. Section R110.4 allows the building official to issue a temporary certificate of occupancy, provided that such portion or portions shall be occupied safely. Although this appears to address life safety items, such as structural integrity, stairs, handrails, means of egress, and smoke alarms, the building official has the authority to require that all requirements of the code are met at the time of the final inspection. Exterior weather barrier materials, such as roofing and siding, may not seem to be directly connected to "occupied safely," but if there is significant water intrusion into the dwelling that causes mold or deterioration of the structure, then it will be an enormous task to try to correct after the house is occupied.

Q: Can landscaping items be required prior to the issuance of a certificate of occupancy?

A: In a situation where the final inspection of the dwelling has been completed and approved by the building official, yet items such as landscaping or hard-surface driveways that are not required by the code, but may be required by a local zoning ordinance have not been completed, the building official has the authority to issue a certificate of occupancy for the structure. The landscape issues, such as sod, trees or a hard-surface driveway can be addressed per the rules of the municipality.

NOTES

DEFINITIONS

CHAPTER

2

SECTION R201
GENERAL

R201.3 Terms defined in other codes. Where terms are not defined in this code such terms shall have meanings ascribed to them as in other code publications of the International Code Council.

ACCESSORY STRUCTURE. A structure not greater than 3,000 square feet (279 m^2) in floor area, and not over two stories in height, the use of which is customarily accessory to and incidental to that of the dwelling(s) and which is located on the same lot..

Q: We understand that an accessory structure is limited to 3,000 square feet by definition. Are there any provisions that would permit accessory space greater than 3,000 square feet, such as 6,000 square feet, if needed?

A: Yes. Provided the local zoning ordinance was not more restrictive, two accessory structures, each not exceeding 3,000 square feet, could be constructed on the same lot. If the distance between the exterior walls of the structures is less than 10 feet, or if the structures are attached, an imaginary lot line is created between the structures. The exterior walls facing the imaginary line would be required to meet the fire-resistant protection requirements of Section R302.1. See the definition of "Fire separation distance" and the related discussion of Section R302.1 in Chapter 3.

Q: A development plan has been submitted that contains two long rows of single-family residential dwellings on individual lots. Splitting these two long rows is an aircraft runway that will be used to serve these dwellings. The plan is to construct a private aircraft hangar on each individual lot, so the owner of the dwelling can land his or her aircraft, park it in his or her own private hangar and walk a few steps into the dwelling. Can a private aircraft hangar be considered an accessory structure to a residential dwelling?

A: Yes. The aircraft hangar in the planned development that you describe could be considered an accessory structure of the dwelling.

APPROVED. Acceptable to the building official.

Q: At one time the code made reference to tests that a building official needed in order to approve something. Are those tests no longer required?

A: Testing is not required, but is often used to verify compliance with a particular code or standard. The building official will make the determination on what submittal documents or testing is needed.

BALCONY, EXTERIOR. An exterior floor projecting from and supported by a structure without additional independent supports.

DECK. An exterior floor system supported on at least two opposing sides by an adjoining structure and/or posts, piers, or other independent supports.

Q: What is the difference between a deck and exterior balcony?

A: The floor system of a deck is supported on at least two sides, where the floor system of a balcony is projected from a structure without additional independent supports. It is a cantilevered floor. See Figure 2-1.

NONCOMBUSTIBLE MATERIAL. Materials that pass the test procedure for defining noncombustibility of elementary materials set forth in ASTM E 136.

Q: When the code makes reference to "noncombustible materials," such as when it discusses clearances around masonry chimneys, who determines if the material is noncombustible?

A: The building official makes that determination based on the submittal documents. To be considered noncombustible, the material would need to be tested to ASTM E 136, *Standard Test Method For Behavior Of Materials in a Vertical Tube Furnace At 750 °C.* This standard is used to measure and describe the response of materials, products or assemblies to heat and flame under controlled conditions. Since 1920, a series of building and fire codes have changed the definition of what is considered noncombustible. ASTM E 136 was promulgated as a full standard in 1965, and used with slight changes since that time.

BALCONY
FIGURE 2-1

RUNNING BOND. The placement of masonry units such that head joints in successive courses are horizontally offset at least one-quarter the unit length.

STACK BOND. The placement of masonry units in a bond pattern is such that head joints in successive courses are vertically aligned. For the purpose of this code, requirements for stack bond shall apply to all masonry laid in other than running bond.

Q: Why is it important to know the difference between running bond and stack bond in the code?

A: There are provisions in the code related to the use of stack bond masonry that do not apply to masonry laid in a running bond. For example, Section R606.8 requires longitudinal reinforcement consisting of not less than two continuous wires, each with a minimum aggregate cross-sectional area of 0.017 square inches (9 US gage) in horizontal bed joints spaced not more than 16 inches on center vertically for unreinforced masonry walls laid in stack bond.

TOWNHOUSE. A single-family dwelling unit constructed in a group of three or more attached units in which each unit extends from foundation to roof and with open space on at least two sides.

Q: What is a townhouse?

A: A townhouse, defined in Section R202, is a group of three or more attached units in which each unit extends from the foundation to roof and with open space on at least two sides. See Figure 2-2. If the proposed structure does not comply with all of the three provisions in the definition, it is not a townhouse. For example, six dwelling units back-to-back would not be considered a townhouse because the middle two units would not have two sides open.

TOWNHOUSE
FIGURE 2-2

Q: What is meant by open space on at least two sides?

A: Figure 2-3 shows a three-unit townhouse. Unit A and Unit C are open front and back, and on the ends, for a total of three sides open. Unit B is open front and back, for a total of two sides open. "Open" means that the entire side adjoins a yard or public space.

Q: Are four dwelling units back-to-back without lot lines between each unit considered a townhouse by definition?

A: Yes, each unit is open on two sides. See Figure 2-4.

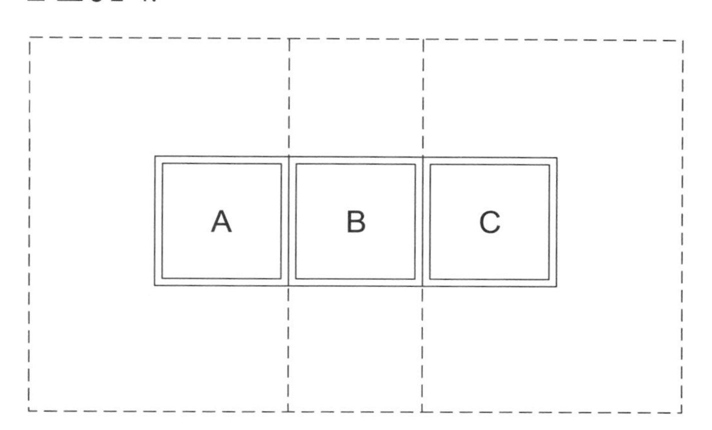

TOWNHOUSE UNITS OPEN ON TWO SIDES
FIGURE 2-3

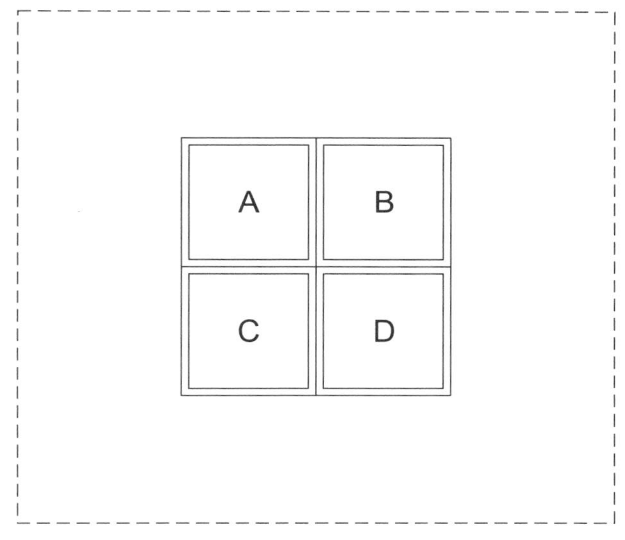

FOUR-UNIT TOWNHOUSE WITHOUT LOT-LINE SEPARATION
FIGURE 2-4

Q: Does a four-dwelling unit back-to-back townhouse with lot lines between each dwelling unit change any requirements based on adding the lot lines?

A: No. Fire separation distance is evaluated the same with or without lot lines. See Figure 2-5.

Q: Will the five dwelling units in Figure 2-6 meet the definition of "Townhouse"?

A: Yes. All units have at least two sides open.

Q: Will the six dwelling units back-to-back in Figure 2-7 meet the definition of "Townhouse"?

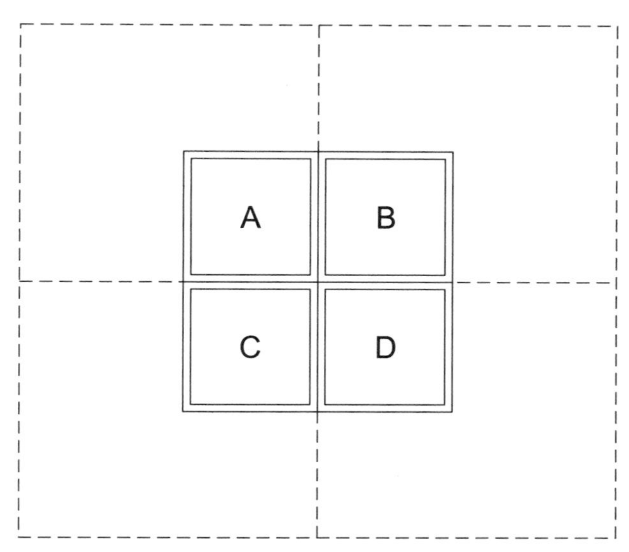

FOUR-UNIT TOWNHOUSE WITH LOT-LINE SEPARATION
FIGURE 2-5

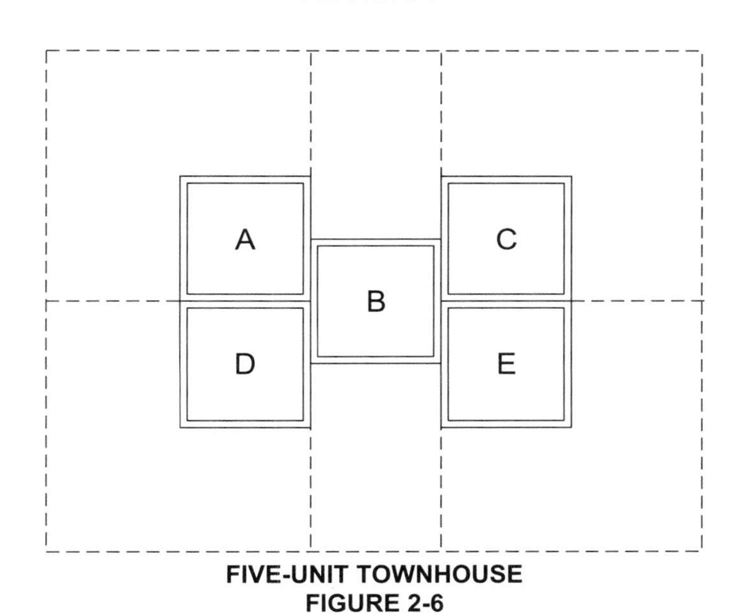

FIVE-UNIT TOWNHOUSE
FIGURE 2-6

A: No. Unit B and Unit E are not open on two sides.

Q: Could the six dwelling units back-to-back in the drawing for the previous question be constructed using some other provision in the code?

A: No. Construction of this building must comply with the IBC.

Q: What is the maximum number of dwelling units permitted in a townhouse?

A: There is no limit to the number of dwelling units.

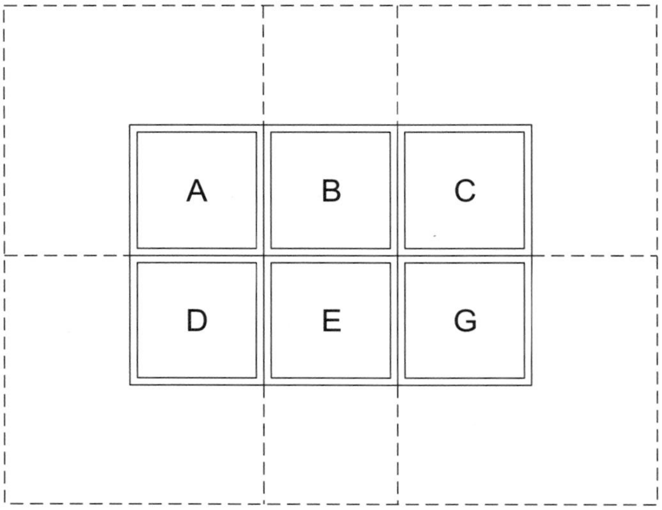

SIX-UNIT STRUCTURE
FIGURE 2-7

Q: Can a townhouse contain stacked units (upper and lower dwelling units)?

A: No. The definition for townhouses only allows single units that are completely separated from foundation to roof sheathing.

Q: For a townhouse unit to be considered open, such as in dwelling Unit B and dwelling Unit E in Figure 2-8, does the entire side need to be open for the full length of that side of that dwelling unit?

A: Yes. The definition says open on two sides. There is nothing to indicate a portion or percentage of a side is acceptable. Dwelling Units B and E do not have two sides open.

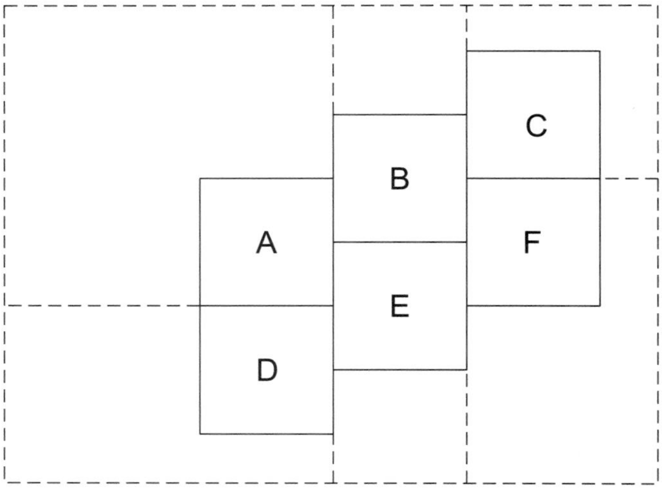

SIX-UNIT STRUCTURE
FIGURE 2-8

NOTES ______________________________

CHAPTER

BUILDING PLANNING

3

SECTION R301 DESIGN CRITERIA

R301.1 Application. Buildings and structures, and all parts thereof, shall be constructed to safely support all loads, including dead loads, live loads, roof loads, flood loads, snow loads, wind loads and seismic loads as prescribed by this code. The construction of buildings and structures in accordance with the provisions of this code shall result in a system that provides a complete load path that meets all requirements for the transfer of all loads from their point of origin through the load-resisting elements to the foundation. Buildings and structures constructed as prescribed by this code are deemed to comply with the requirements of this section.

R301.1.2 Construction systems. The requirements of this code are based on platform and balloon-frame construction for light-frame buildings. The requirements for concrete and masonry buildings are based on a balloon framing system. Other framing systems must have equivalent detailing to ensure force transfer, continuity and compatible deformations.

Q: Do buildings constructed under the IRC require engineering?

A: Not necessarily. The code provides prescriptive information that can be used to completely build a dwelling without using the services of a design professional. For instance, the tables in the code for spans of floor joists and heights of walls are based on the use of grade-stamped solid sawn materials that allow the user to "plug-in" his or her particular design loads and find the required lumber. The wall and floor sheathing tables are based on graded sheathing that allows the use of standardized material for this conventional light framed construction project. The concrete and masonry wall tables allow the user to design a wall based on prescriptive information such as known wall heights, proposed wall thicknesses, weights of unbalanced fills and strength of concrete to determine reinforcing that may be required. When dwellings are constructed with wall heights that exceed the tables in the code, floors that exceed the limits in the span tables and bearing points that are cantilevered at many locations, then the design of the structure is exceeding the limits of the prescriptive information in the code.

R301.1.3 Engineered design. When a building of otherwise conventional construction contains structural elements exceeding the limits of Section R301 or otherwise not conforming to this code, these elements shall be designed in accordance with accepted engineering practice. The extent of such design need only demonstrate compliance of nonconventional elements with other applicable provisions and shall be compatible with the performance of the conventional framed system. Engineered design in accordance with the *International Building Code* is permitted for all buildings and structures, and parts thereof, included in the scope of this code.

Q: When designing trusses for dwellings and accessory structures in our northern climate, should the truss designer use ground snow loads or roof snow loads?

A: Engineered trusses are designed for roof snow loads. However, the ground snow load of the geographic region is the basis for determining the roof snow load design criteria. The building official must verify that the correct design criteria are used and indicated on the truss design drawings.

Q: I understand that the IBC requires trusses to be evaluated for unbalanced snow loading from all directions, but the IRC does not appear to require this same evaluation. Is this correct?

A: Section R802.10.2 requires wood trusses to be designed in accordance with accepted engineering practice. This means the trusses must be designed according to the IBC, which references ASCE 7, National Design Specification for Wood Construction (NDS) and Truss Plate Institute (TPI) standards. ASCE 7 requires the truss engineer to consider unbalanced loading as well as other possible loading conditions depending on the configuration of the roof.

Q: When the design of a dwelling appears to contain structural elements that are not addressed in the code, Section R301.1.3 requires those elements to be designed in accordance with accepted engineering practice. What is accepted engineering practice?

A: Although not defined in the code, accepted engineering practice means the engineering analysis is based on well-established principles of mechanics and conforms to accepted principles, tests or standards of nationally recognized technical or scientific authorities.

Q: Does the code require that this information be provided by a design professional, such as an architect or engineer?

A: Yes. When the code requires engineering it means the design must be prepared by a registered design professional as defined in Section R202. However, the laws of the state or jurisdiction regulate engineering practice and what specific type of professional license is required.

R301.4 Dead load. The actual weights of materials and construction shall be used for determining dead load with consideration for the dead load of fixed service equipment.

Q: Are furniture, bookcases and cabinetry considered dead loads?

A: The dead load of a dwelling would include the actual weight of the walls, floor, roof, ceiling, stairways, fixed service equipment and other permanent materials of construction incorporated into the building, such as finishes and architectural items. Built-in cabinetry and bookcases could be considered dead load; furniture would be considered live load.

R301.5 Live load. The minimum uniformly distributed live load shall be as provided in Table R301.5.

Q: Is the 50-pounds-per-square-foot (psf) live load for passenger vehicle garages added to the 2,000 pound live load over a 20-square-inch area?

A: The garage floor needs to be designed to carry both loads. The 50 psf live load along with the dead load of the building materials will be the major considerations for the design of the foundation wall and footing sizes. The 50-psf uniform load and the 2,000-pound concentrated load do not act concurrently. See Figure 3-1.

TABLE R301.5
MINIMUM UNIFORMLY DISTRIBUTED LIVE LOADS
(in pounds per square foot)

USE	LIVE LOAD
Attics with limited storage[b, g, h]	20
Attics without storage[b]	10
Decks[e]	40
Exterior balconies	60
Fire escapes	40
Guardrails and handrails[d]	200[i]
Guardrails in–fill components[f]	50[i]
Passenger vehicle garages[a]	50[a]
Rooms other than sleeping rooms	40
Sleeping rooms	30
Stairs	40[c]

For SI: 1 pound per square foot = 0.0479 kPa, 1 square inch = 645 mm^2, 1 pound = 4.45 N.

a. Elevated garage floors shall be capable of supporting a 2,000-pound load applied over a 20-square-inch area.

b. Attics without storage are those where the maximum clear height between joist and rafter is less than 42 inches, or where there are not two or more adjacent trusses with the same web configuration capable of containing a rectangle 42 inches high by 2 feet wide, or greater, located within the plane of the truss. For attics without storage, this live load need not be assumed to act concurrently with any other live load requirements.

c. Individual stair treads shall be designed for the uniformly distributed live load or a 300-pound concentrated load acting over an area of 4 square inches, whichever produces the greater stresses.

d. A single concentrated load applied in any direction at any point along the top.

e. See Section R502.2.1 for decks attached to exterior walls.

f. Guard in-fill components (all those except the handrail), balusters and panel fillers shall be designed to withstand a horizontally applied normal load of 50 pounds on an area equal to 1 square foot. This load need not be assumed to act concurrently with any other live load requirement.

g. For attics with limited storage and constructed with trusses, this live load need be applied only to those portions of the bottom chord where there are two or more adjacent trusses with the same web configuration capable of containing a rectangle 42 inches high or greater by 2 feet wide or greater, located within the plane of the truss. The rectangle shall fit between the top of the bottom chord and the bottom of any other truss member, provided that each of the following criteria is met:

1. The attic area is accessible by a pull-down stairway or framed opening in accordance with Section R807.1; and
2. The truss has a bottom chord pitch less than 2:12.

h. Attic spaces served by a fixed stair shall be designed to support the minimum live load specified for sleeping rooms.

i. Glazing used in handrail assemblies and guards shall be designed with a safety factor of 4. The safety factor shall be applied to each of the concentrated loads applied to the top of the rail, and to the load on the in-fill components. These loads shall be determined independent of one another, and loads are assumed not to occur with any other live load.

Q: Where do I find the loading requirement for a handrail on a residential stairway?

A: Guardrails and handrails need to be designed for a single concentrated load of 200 pounds applied in any direction at any point along the top of the rail per Table R301.5.

PRE-CAST CONCRETE FOR GARAGE FLOOR
FIGURE 3-1

Q: Does this 200-pound load include any required guardrail components?

A: Only the top rail of the guard must be designed to resist the concentrated 200-pound load. The guard in-fill components shall be designed to withstand a horizontally applied normal load of 50 pounds spread over an area equal to 1 square foot per Note f in Table R301.5.

Q: We have read Note i in Table R301.5. What is the reference to a safety factor of 4?

A: The design live loads for glazing used in handrail assemblies need to be multiplied by four. Also, all glazing in guards needs to comply with Section R308.4 for safety glazing. See Figure 3-2.

Q: Does the code require glazing used in a guard to be supported on all sides?

A: No. The manufacturer of the glazing will need to provide the design data, along with the installation requirements for the guard, which may require the glazing to be supported on all sides.

GLAZING COMPONENT IN GUARD ON DECK
FIGURE 3-2

Q: Is a deck stair designed for the same live load as the deck it serves?

A: The deck surface is designed for a live load of 40 psf per Table R301.5. Note c requires that a deck stair be designed for a live load of 40 psf or a 300-pound concentrated load acting over an area of 4 square inches, whichever produces the greater load. See Figure 3-3.

DECK STAIRWAY
FIGURE 3-3

Q: A plan has been submitted for a permit that indicates an attached garage with attic storage trusses above. A code-complying stairway is proposed for access to the storage area in the attic. What design load is required for this attic storage area?

A: When accessed by a fixed stair, Note h of Table R301.5 requires the attic space to be designed to support the minimum live load specified for sleeping rooms (30 psf). Per Note b, the live load for the attic storage trusses needs to be applied only to those portions of the bottom chord where there are two or more adjacent trusses with the same web configuration capable of containing a rectangle 42 inches high or greater by 2 feet wide or more, located within the plane of the truss. The submittal documents should be reviewed to determine if the attic storage area will need to be designed for greater than 30 psf based upon the intended storage capacity. If a greater capacity is anticipated, it must be reflected on the engineered truss drawings.

Q: Sleeping rooms are designed for 30 psf live load. What live load is required for a closet in the sleeping room, or an attached bathroom?

A: Sleeping rooms need to be designed for 30 psf live load. Rooms other than sleeping rooms must be designed for 40-psf live load. It is reasonable for the closet and attached bathroom to be designed for the same loading as the sleeping room that they serve, which is 30 psf.

R301.6 Roof load. The roof shall be designed for the live load indicated in Table R301.6 or the snow load indicated in Table R301.2(1), whichever is greater.

TABLE R301.6
MINIMUM ROOF LIVE LOADS IN POUNDS-FORCE PER SQUARE FOOT OF HORIZONTAL PROJECTION

ROOF SLOPE	TRIBUTARY LOADED AREA IN SQUARE FEET FOR ANY STRUCTURAL MEMBER		
	0 to 200	201 to 600	Over 600
Flat or rise less than 4 inches per foot (1:3)	20	16	12
Rise 4 inches per foot (1:3) to less than 12 inches per foot (1:1)	16	14	12
Rise 12 inches per foot (1:1) and greater	12	12	12

For SI: 1 square foot = 0.0929 m^2, 1 pound per square foot = 0.0479 kPa, 1 inch per foot = 83.3 mm/m.

Q: How do we determine the appropriate snow loads for our area?

A: Refer to the ground snow load map in Figure R301.2(5). In some regions, (indicated by "CS" on the map), the ground snow load must be determined by site-specific case studies based on the requirements of ASCE 7. This could also be determined locally by state or municipal authority based upon historical evidence without referencing the map.

R301.7 Deflection. The allowable deflection of any structural member under the live load listed in Sections R301.5 and R301.6 shall not exceed the values in Table R301.7.

TABLE R301.7
ALLOWABLE DEFLECTION OF STRUCTURAL MEMBERS[a,b,c]

STRUCTURAL MEMBER	ALLOWABLE DEFLECTION
Rafters having slopes greater than 3/12 with no finished ceiling attached to rafters	L/180
Interior walls and partitions	H/180
Floors and plastered ceilings	L/360
All other structural members	L/240
Exterior walls with plaster or stucco finish	H/360
Exterior walls—wind loads[a] with brittle finishes	L/240
Exterior walls—wind loads[a] with flexible finishes	L/120

Note: L = span length, H = span height.

a. The wind load shall be permitted to be taken as 0.7 times the Component and Cladding loads for the purpose of the determining deflection limits herein.

b. For cantilever members, L shall be taken as twice the length of the cantilever.

c. For aluminum structural members or panels used in roofs or walls of sunroom additions or patio covers, not supporting edge of glass or sandwich panels, the total load deflection shall not exceed L /60. For sandwich panels used in roofs or walls of sunroom additions or patio covers, the total load deflection shall not exceed L/120.

Q: We have a project that has a second floor wood joist system with a plaster ceiling applied to the bottom side. Could you explain what the L/360 allowable deflection refers to based on a joist system with a 12-foot span?

A: "L" is the span length. For your example of a 12-foot span (L= 144 inches), the span divided by 360 equals the maximum allowable deflection of 0.4 inches.

R301.8 Nominal sizes. For the purposes of this code, where dimensions of lumber are specified, they shall be deemed to be nominal dimensions unless specifically designated as actual dimensions.

Q: We have always considered a 2 x 4 to be a nominal size of 1½ inches by 3½ inches. Where does that determination come from?

A: The nominal size is 2 inches by 4 inches and the actual size is approximately 1½ inches by 3½ inches. Nominal lumber size is the commercial size designation of width and depth in standard sawn lumber that is somewhat larger than the standard net size of dressed lumber in accordance with U.S. Department of Commerce PS-20. Refer to Section 2302 of the IBC for a definition of "Nominal" lumber size. Nominal lumber sizes are contained in the NDS.

SECTION R302
EXTERIOR WALL LOCATION

R302.1 Exterior walls. Construction, projections, openings and penetrations of exterior walls of dwellings and accessory buildings shall comply with Table R302.1. These provisions shall not apply to walls, projections, openings or penetrations in walls that are perpendicular to the line used to determine the fire separation distance. Projections beyond the exterior wall shall not extend more than 12 inches (305 mm) into the areas where openings are prohibited.

Exceptions:

1. Detached tool sheds and storage sheds, playhouses and similar structures exempted from permits are not required to provide wall protection based on location on the lot. Projections beyond the exterior wall shall not extend over the lot line.
2. Detached garages accessory to a dwelling located within 2 feet (610 mm) of a lot line are permitted to have roof eave projections not exceeding 4 inches (102 mm).
3. Foundation vents installed in compliance with this code are permitted.

Q: Table R302.1 (below) states that 1-hour fire-resistance-rated walls (with exposure from both sides) can have a minimum fire separation distance of zero feet. Is this generally the basis for why townhouses have two 1-hour fire-resistance-rated walls between each unit?

A: Yes. Refer to the Q&A for Section R321 for examples.

Q: The exterior wall of a dwelling is located 5 feet from a lot line. Is this exterior wall required to be fire-resistance rated?

A: No. See Table R302.1, "Exterior Walls." If the exterior wall is less than 5 feet from the lot line, it needs to be constructed as a 1-hour fire-resistance-rated wall with exposure from both sides. See Figure 3-4.

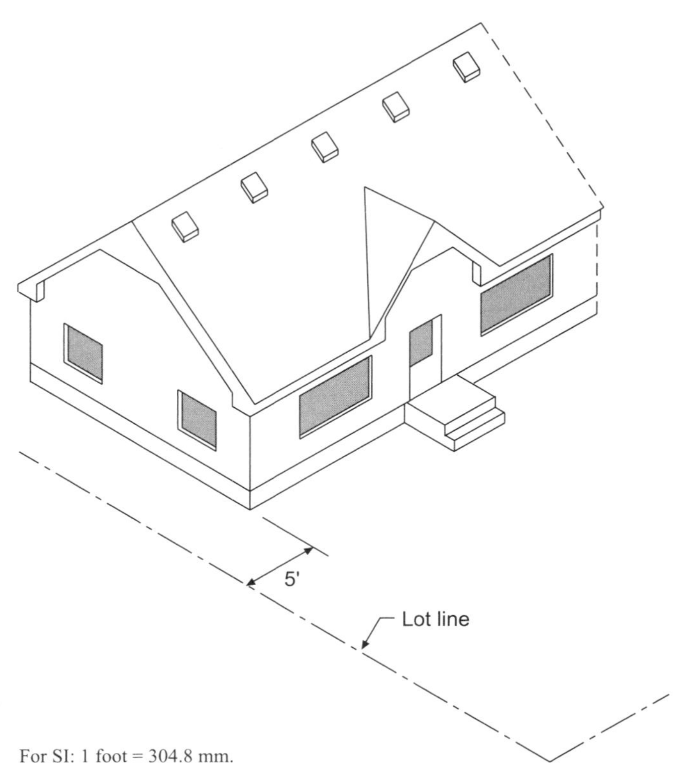

For SI: 1 foot = 304.8 mm.

DWELLING 5 FEET FROM LOT LINE
FIGURE 3-4

Q: For a dwelling 5 feet from the lot line, is the 12-inch gable end roof overhang required to be fire-resistance rated?

A: Yes. Roof overhang projections less than 5 feet from the lot line require a 1-hour fire-resistance rating on the underside of the overhang in accordance with Table R302.1.

Q: The gable end exterior wall of a dwelling is located 4 feet from a lot line. Is it required to fire-resistance rated?

A: Yes. Table R302.1, "Exterior Walls," would require this gable end exterior wall that is located parallel to the lot line, and 4 feet away, to be 1-hour fire-resistance rated (with exposure from both sides). The provisions in the code do not apply to the perpendicular walls of the structure in the drawing. Fire separation distance, referred to in Section R302.1, is measured perpendicular to the face of the exterior wall. Note that for dwellings not parallel to the lot line, the fire separation distance varies along the length of the wall, and only portions of the wall measuring less than 5 feet to the lot line require the fire-resistance rating. See Figure 3-5.

TABLE R302.1
EXTERIOR WALLS

EXTERIOR WALL ELEMENT		MINIMUM FIRE-RESISTANCE RATING	MINIMUM FIRE SEPARATION DISTANCE
Walls	(Fire-resistance rated)	1 hour with exposure from both sides	0 feet
	(Not fire-resistance rated)	0 hours	5 feet
Projections	(Fire-resistance rated)	1 hour on the underside	2 feet
	(Not fire-resistance rated)	0 hours	5 feet
Openings	Not allowed	N/A	< 3 feet
	25% Maximum of Wall Area	0 hours	3 feet
	Unlimited	0 hours	5 feet
Penetrations	All	Comply with Section R317.3	< 5 feet
		None required	5 feet

N/A = Not Applicable.

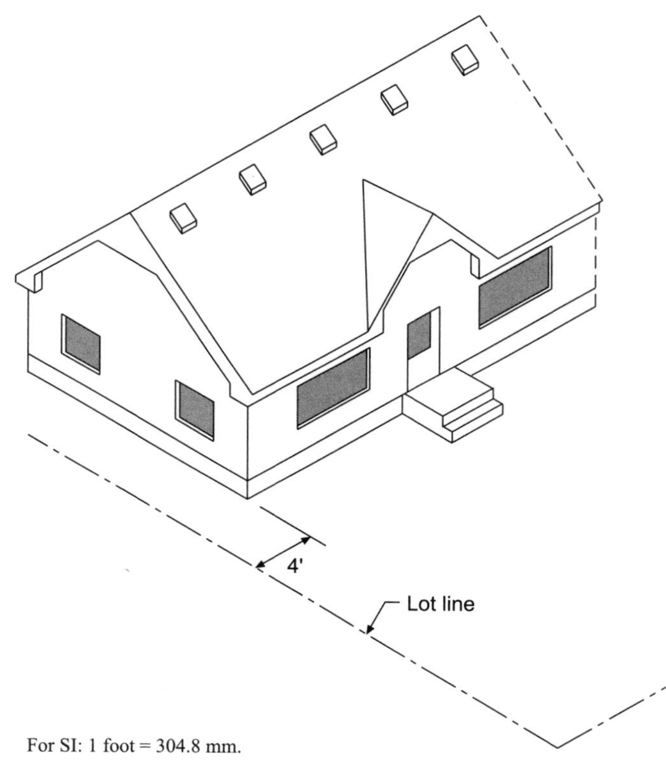

For SI: 1 foot = 304.8 mm.

DWELLING 4 FEET FROM LOT LINE
FIGURE 3-5

Q: Are the windows in the gable end wall permitted?

A: The windows are permitted openings in this gable end exterior wall as long as they do not exceed 25 percent of the wall area. For example, if the gable end wall covers 240 square feet, the window or door openings (combined) could not exceed 60 square feet.

Q: With the exterior wall 4 feet from the lot line, could the area of these windows be increased if they were designed with fire protection-rated glazing?

A: Section R302.1 allows 25-percent openings, whether they are rated or not, and has no provisions for any increase over 25 percent.

Q: Assuming the gable end exterior wall located 4 feet from the lot line is 1-hour fire-resistance rated, what protection is needed for a 12-inch gable end overhang?

A: The gable end overhang, extending to 3 feet from the lot line, requires 1-hour fire resistance on the underside of the overhang.

Q: Is the fascia board of this 12-inch overhang required to be protected?

A: No. Table R302.1 only requires 1-hour protection on the underside of the projection.

Q: With the gable end exterior wall located 4 feet from the lot line constructed as a 1-hour fire-resistance-rated wall, are the front or rear walls of the dwelling required to be fire-resistance rated?

A: No. The windows in the front and rear wall can also be unprotected and of unlimited area. The provisions in Section R302.1 shall not apply to walls, projections, openings or penetrations in walls that are perpendicular to the line used to determine the fire separation distance.

Q: If a 1-hour fire-resistance-rated gable end exterior wall is located 4 feet from the lot line, what is the maximum projection allowed for a gable end roof overhang with 1-hour fire-resistance rating on the bottom side of the projection?

A: Section R302.1 limits projections beyond the exterior wall to not extend more than 12 inches in the area where openings are prohibited. Openings are prohibited less than 3 feet from the lot line. This would allow the roof overhang to extend to a point 2 feet from the lot line, or with the dwelling exterior wall 4 feet from the lot line, the maximum roof overhang is 2 feet.

Q: Two dwelling units are constructed side by side with a lot line running between the two units. What level of fire-resistance rating is required for this situation?

A: The two dwelling units would essentially be treated the same as two separate dwelling units located on the lot line. Because of the lot line between the two dwelling units, the code would require two 1-hour fire-resistance-rated walls between the two units. See Figure 3-6.

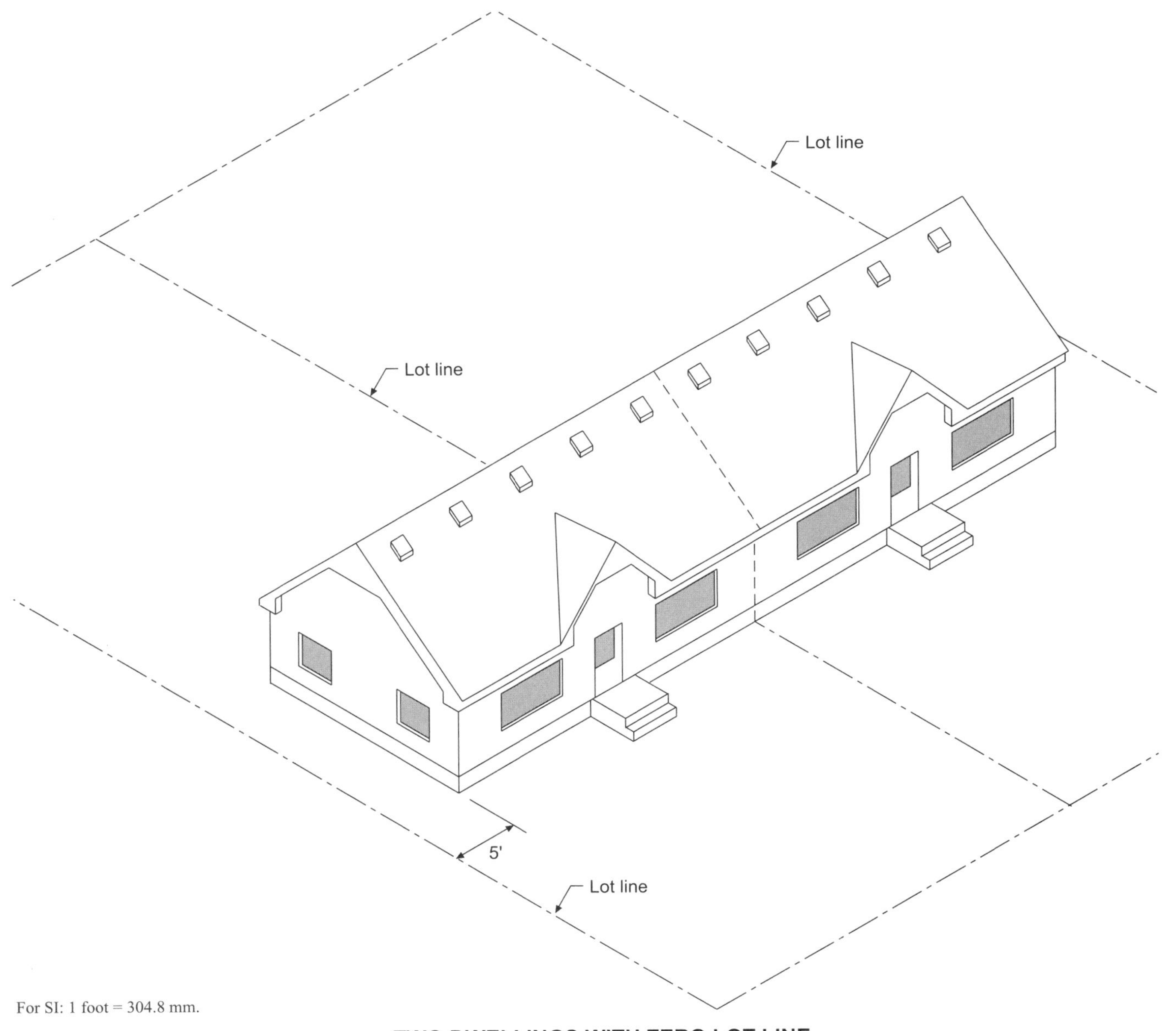

For SI: 1 foot = 304.8 mm.

TWO DWELLINGS WITH ZERO LOT LINE
FIGURE 3-6

Q: How close to a lot line can a 100-square-foot detached tool shed be located?

A: If this tool shed did not have any roof overhang, it could be placed right at the lot line. See Figure 3-7.

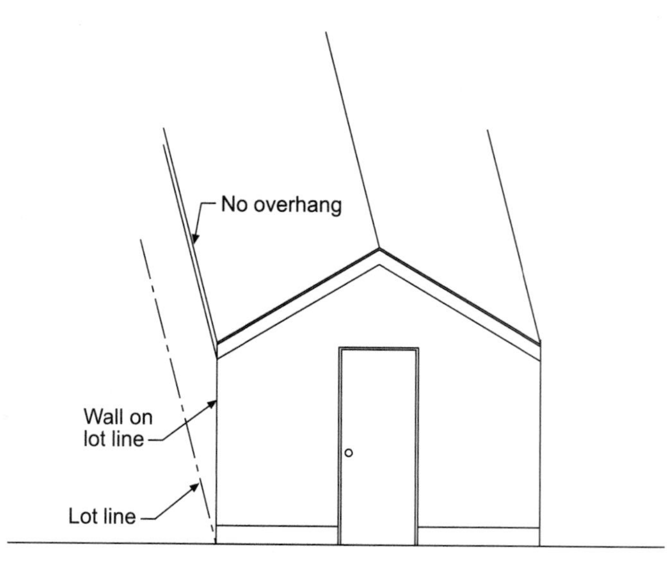

ACCESSORY STRUCTURE ON LOT LINE
FIGURE 3-7

Q: What if this shed had a roof overhang of 1 foot, 6 inches?

A: If the tool shed had a roof overhang of 1 foot, 6 inches, it could be placed 1 foot, 6 inches from the lot line. See Figure 3-8.

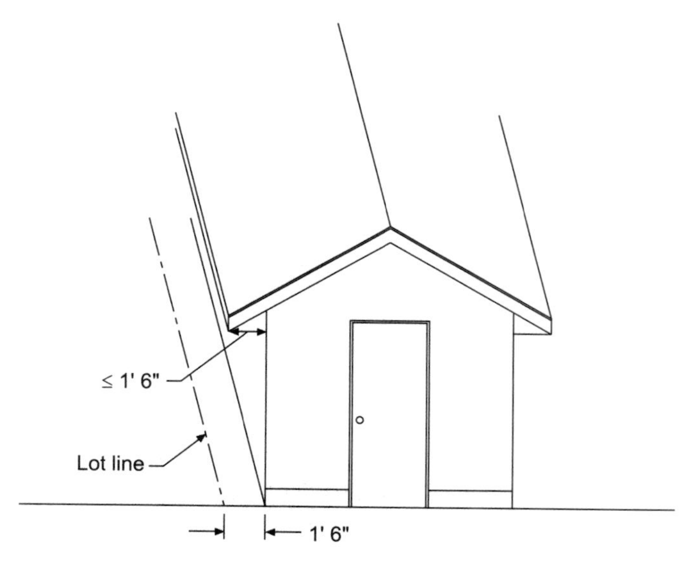

For SI: 1 inch = 25.4 mm, 1 foot = 304.8 mm.

ACCESSORY STRUCTURE 1 FOOT, 6 INCHES FROM LOT LINE
FIGURE 3-8

Q: Are window openings permitted in the exterior wall of this tool shed with the exterior wall 1 foot 6 inches from the lot line?

A: Because this tool shed is exempt from permits, it is also exempt from all fire-resistance rated requirements based on location on the lot under Section R302.1. The wall can be constructed of any materials permitted by the code. There is no area limitation on the unprotected window openings.

Q: How close can the nonfire-resistance-rated exterior wall of a 200-square-foot detached garage be located to a lot line?

A: The provisions in Table R302.1 for exterior walls apply to this structure. If the exterior wall of this detached garage is located at least 5 feet away from a lot line, there is no requirement for a fire-resistance-rated wall. See Figure 3-9.

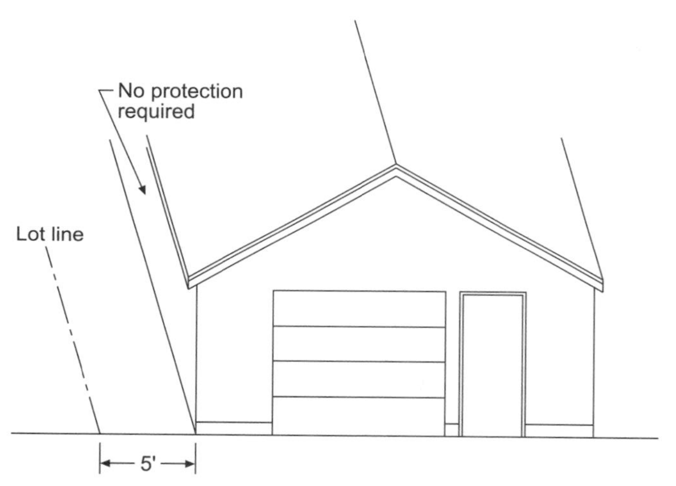

For SI: 1 foot = 304.8 mm.

200 SQ. FT. DETACHED GARAGE ADJACENT TO LOT LINE
FIGURE 3-9

Q: If a detached garage is located 1 foot, 6 inches from the lot line, what is the maximum roof overhang toward the lot line?

A: In Exception 2 of Section R302.1, the code would limit the roof overhang, or projection, to 4 inches. See Figure 3-10.

Q: Why is there a difference in the roof eave restriction between the shed and the detached garage?

A: The larger building could have a higher fuel load potential.

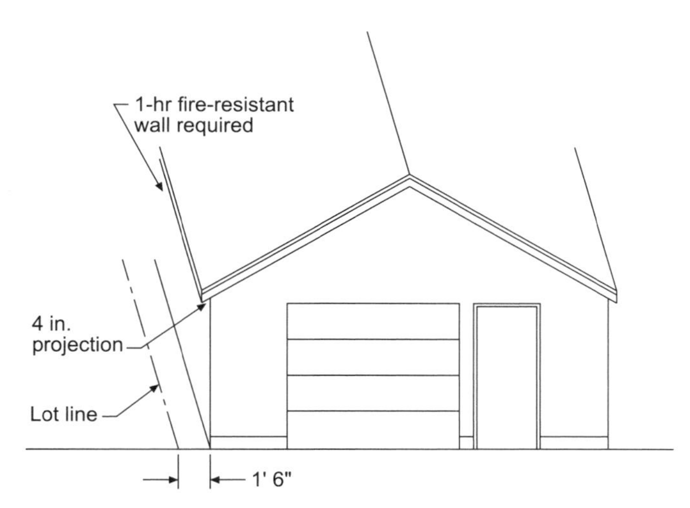

For SI: 1 inch = 25.4 mm, 1 foot = 304.8 mm.

DETACHED GARAGE ROOF PROJECTION
FIGURE 3-10

Q: In our city, there are many very old homes that are located within a few feet of the side lot lines. See Figure 3-11. Some of these lots are being rezoned for retail use, and often there is a situation where a retail occupancy is constructed right on the lot line. Are there any code provisions that would require the neighboring homeowner to upgrade the fire resistance of his or her exterior wall based on the location of the new retail occupancy a couple of feet away?

A: No.

EXISTING DWELLINGS ADJACENT TO LOT LINE
FIGURE 3-11

SECTION R303
LIGHT, VENTILATION AND HEATING

R303.1 Habitable rooms. All habitable rooms shall have an aggregate glazing area of not less than 8 percent of the floor area of such rooms. Natural ventilation shall be through windows, doors, louvers or other approved openings to the outdoor air. Such openings shall be provided with ready access or shall otherwise be readily controllable by the building occupants. The minimum openable area to the outdoors shall be 4 percent of the floor area being ventilated.

Exceptions:

1. The glazed areas need not be openable where the opening is not required by Section R310 and an approved mechanical ventilation system capable of producing 0.35 air change per hour in the room is installed or a whole-house mechanical ventilation system is installed capable of supplying outdoor ventilation air of 15 cubic feet per minute (cfm) (78 L/s) per occupant computed on the basis of two occupants for the first bedroom and one occupant for each additional bedroom.
2. The glazed areas need not be installed in rooms where Exception 1 above is satisfied and artificial light is provided capable of producing an average illumination of 6 footcandles (65 lux) over the area of the room at a height of 30 inches (762 mm) above the floor level.
3. Use of sunroom additions and patio covers, as defined in Section R202, shall be permitted for natural ventilation if in excess of 40 percent of the exterior sunroom walls are open, or are enclosed only by insect screening.

Q: Can the minimum ventilation requirement for habitable rooms be met with a combination of natural ventilation and mechanical ventilation?

A: The code has no provisions for combining natural and mechanical ventilation in order to achieve the required ventilation.

Q: Could the building official approve a combination of natural ventilation and mechanical ventilation as an alternative to the code?

A: Yes. The building official would need to evaluate the effectiveness of the design, and whether it ultimately delivers the minimum ventilation rate.

Q: For the installation of a bathroom exhaust fan, the heating, ventilating and air-conditioning (HVAC) and plumbing contractors in our area ask, what is considered "directly to the outside"?

A: In Section R303.3, "directly to the outside" intends that the vent extends through the roof, wall or soffit to the outside air. For example, an exhaust vent incorrectly terminated inside of an attic space is not considered directly to the outside. Such an installation could create moisture problems (condensation) and possibly health and structural problems.

Q: Could the vent terminate on the bottom face of a roof jack or ridge vent?

A: No.

Q: Could the vent terminate on the bottom side of a soffit?

A: It would need to have an approved termination fitting so air is not exhausted into the soffit.

Q: Would the code require two bath fans for a bathroom with a separated toilet compartment off the area with the shower and/or bathtub?

A: Yes. A separate exhaust fan or openable window is required for each enclosed space.

R303.4 Opening location. Outdoor intake and exhaust openings shall be located in accordance with Sections R303.4.1 and R303.4.2.

R303.4.1 Intake openings. Mechanical and gravity outdoor air intake openings shall be located a minimum of 10 feet (3048 mm) from any hazardous or noxious contaminant, such as vents, chimneys, plumbing vents, streets, alleys, parking lots and loading docks, except as otherwise specified in this code. Where a source of contaminant is located within 10 feet (3048 mm) of an intake opening, such opening shall be located a minimum of 2 feet (610 mm) below the contaminant source.

For the purpose of this section, the exhaust from dwelling unit toilet rooms, bathrooms and kitchens shall not be considered as hazardous or noxious.

R303.4.2 Exhaust openings. Outside exhaust openings shall be located so as not to create a nuisance. Exhaust air shall not be directed onto walkways.

Q: The code says, "Outside exhaust openings shall be located so as not to create a nuisance." What does this mean?

A: A nuisance refers to openings located such that the exhaust will be unpleasant or annoying to people on the property or adjacent property.

R303.5 Outside opening protection. Air exhaust and intake openings that terminate outdoors shall be protected with corrosion-resistant screens, louvers or grilles having a minimum opening size of $^1/_4$ inch (6 mm) and a maximum opening size of $^1/_2$ inch (13 mm), in any dimension. Openings shall be protected against local weather conditions. Outdoor air exhaust and intake openings shall meet the provisions for exterior wall opening protectives in accordance with this code.

Q: A new homeowner asked our department if he or she can install $^1/_8$ inch screens on the intakes instead of the ¼ inch screens that are in place because of an issue with bees entering the dwelling through the openings. Is this permitted?

A: No. The code does not permit the reduction in opening. Also, refer to the manufacturer's installation instructions for additional information. See Figure 3-12.

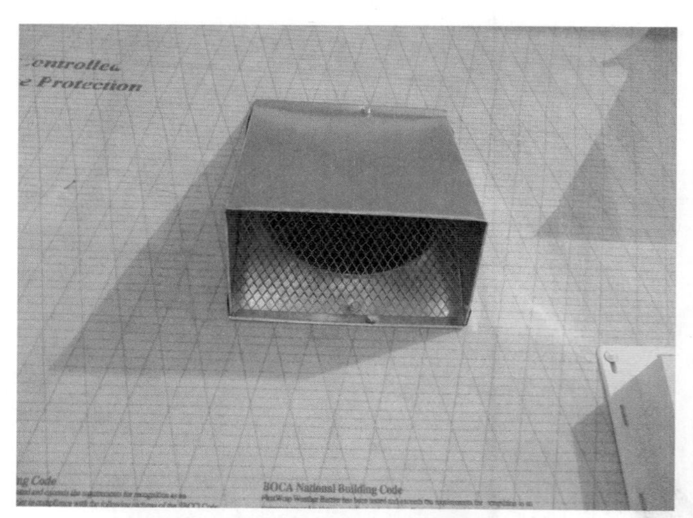

INTAKE OPENING IN EXTERIOR WALL
FIGURE 3-12

R303.6 Stairway illumination. All interior and exterior stairways shall be provided with a means to illuminate the stairs, including the landings and treads. Interior stairways shall be provided with an artificial light source located in the immediate vicinity of each landing of the stairway. For interior stairs the artificial light sources shall be capable of illuminating treads and landings to levels not less than 1 foot-candle (11 lux) measured at the center of treads and landings. Exterior stairways shall be provided with an artificial light source located in the immediate vicinity of the top landing of the stairway. Exterior stairways providing access to a basement from the outside grade level shall be provided with an artificial light source located in the immediate vicinity of the bottom landing of the stairway.

> **Exception:** An artificial light source is not required at the top and bottom landing, provided an artificial light source is located directly over each stairway section.

R303.6.1 Light activation. Where lighting outlets are installed in interior stairways, there shall be a wall switch at each floor level to control the lighting outlet where the stairway has six or more risers. The illumination of exterior stairways shall be controlled from inside the dwelling unit.

> **Exception:** Lights that are continuously illuminated or automatically controlled.

Q: Would a motion detector be considered as "automatically controlled" for exterior lighting of the deck stairs and landing?

A: Yes.

R303.8 Required heating. When the winter design temperature in Table R301.2(1) is below 60°F (16°C), every dwelling unit shall be provided with heating facilities capable of maintaining a minimum room temperature of 68°F (20°C) at a point 3 feet (914 mm) above the floor and 2 feet (610 mm) from exterior walls in all habitable rooms at the design temperature. The installation of one or more portable space heaters shall not be used to achieve compliance with this section.

Q: Section R303.8 states, "capable of maintaining a minimum room temperature of 68°F." We carefully calculate the size of each gas furnace and install it according to code. In our townhouse projects we often have occupants that choose to open the windows during the coldest temperatures just to let in some fresh air, and then complain that the furnace won't heat the dwelling properly. We believe the furnaces are "capable" as long as the occupants don't leave the windows open. Are these furnaces "capable" according to the code language?

A: Yes, provided they are capable of maintaining the required temperature under closed-house conditions. There is no requirement to increase the size of heating equipment to compensate for open windows.

SECTION R304
MINIMUM ROOM AREAS

R304.1 Minimum area. Every dwelling unit shall have at least one habitable room that shall have not less than 120 square feet (11 m^2) of gross floor area.

R304.2 Other rooms. Other habitable rooms shall have a floor area of not less than 70 square feet (6.5 m^2).

Exception: Kitchens.

R304.3 Minimum dimensions. Habitable rooms shall not be less than 7 feet (2134 mm) in any horizontal dimension.

> **Exception:** Kitchens.

R304.4 Height effect on room area. Portions of a room with a sloping ceiling measuring less than 5 feet (1524 mm) or a furred ceiling measuring less than 7 feet (2134 mm) from the finished floor to the finished ceiling shall not be considered as contributing to the minimum required habitable area for that room.

Q: Does the code require at least one sleeping room to be at least 120 square feet of floor area?

A: No. One habitable room, which may be a sleeping room, needs to meet this requirement. Sleeping rooms need to be at least 70 square feet.

Q: It appears that kitchens have no minimum room area or length and width dimensions. Is this correct?

A: Yes.

Q: We have a sleeping room on the upper floor with a sloped ceiling. Can the sloped ceiling height be less than 5 feet?

A: Yes. A sleeping room requires at least 70 square feet of floor area. For this sleeping room with a sloped ceiling, at least 50 percent of the required 70 square feet of floor area needs to be at least 7 feet, 0 inches (35 square feet), and at least 50 percent needs to be at least 5 feet, 0 inches of ceiling height (35 square feet). Once that requirement is met, the code does not address any area with a ceiling height less than 5 feet, 0 inches. See Figure 3-13.

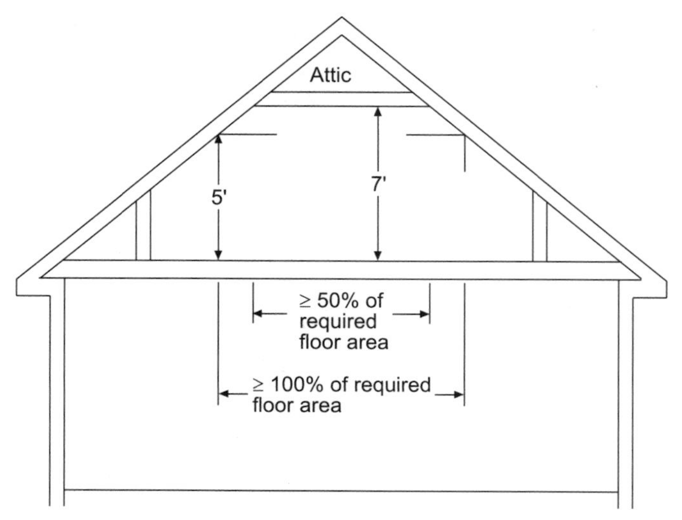

For SI: 1 foot = 304.8 mm.

SLOPED CEILINGS IN HABITABLE SPACE
FIGURE 3-13

SECTION R305 CEILING HEIGHT

R305.1 Minimum height. Habitable rooms, hallways, corridors, bathrooms, toilet rooms, laundry rooms and basements shall have a ceiling height of not less than 7 feet (2134 mm). The required height shall be measured from the finish floor to the lowest projection from the ceiling.

Exceptions:

1. Beams and girders spaced not less than 4 feet (1219 mm) on center may project not more than 6 inches (152 mm) below the required ceiling height.
2. Ceilings in basements without habitable spaces may project to within 6 feet, 8 inches (2032 mm) of the finished floor; and beams, girders, ducts or other obstructions may project to within 6 feet 4 inches (1931 mm) of the finished floor.
3. For rooms with sloped ceilings, at least 50 percent of the required floor area of the room must have a ceiling height of at least 7 feet (2134 mm) and no portion of the required floor area may have a ceiling height of less than 5 feet (1524 mm).
4. Bathrooms shall have a minimum ceiling height of 6 feet 8 inches (2036 mm) over the fixture and at the front clearance area for fixtures as shown in Figure R307.1. A shower or tub equipped with a showerhead shall have a minimum ceiling height of 6 feet 8 inches (2036 mm) above a minimum area 30 inches (762 mm) by 30 inches (762 mm) at the showerhead.

Q: How is ceiling height measured?

A: According to Section R202, ceiling height is considered the clear vertical distance from the finished floor to the finished ceiling.

Q: We read in Section R305.1 that the required height shall be measured from the finished floor to the lowest projection from the ceiling. What are projections?

A: Beams, girders, dropped soffits and ducts would be considered examples of projections.

Q: Are ceiling fans and suspended lighting considered projections? Would the finished ceiling height be measured to the bottom of these fixtures?

A: They are not considered projections. They are fixtures and are not regulated for ceiling height. See Figure 3-14.

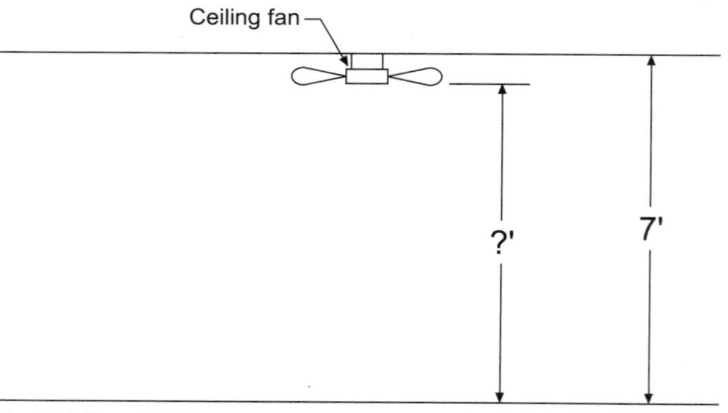

For SI: 1 foot = 304.8 mm.

CEILING HEIGHT DOES NOT APPLY TO CEILING FAN
FIGURE 3-14

Q: When finishing off a basement into living space, such as a recreation room or sleeping room, what is the required minimum clearance under HVAC ducts.

A: 7 feet, 0 inches.

SECTION R306 SANITATION

R306.1 Toilet facilities. Every dwelling unit shall be provided with a water closet, lavatory, and a bathtub or shower.

R306.2 Kitchen. Each dwelling unit shall be provided with a kitchen area and every kitchen area shall be provided with a sink.

Q: Can a single-family dwelling contain a second complete kitchen and separate toilet facilities in the basement level and still be considered a single-family dwelling? Our zoning staff has suggested that we don't issue permits for these basement level finishes that contain these facilities, especially if they have a separate basement level exit door.

A: The IRC does not limit the amount of kitchen or toilet facilities that a single-family dwelling can have in it. Some communities may have zoning regulations that address this situation and enforcement may be assigned to the building department.

SECTION R307 TOILET, BATH AND SHOWER SPACES

R307.1 Space required. Fixtures shall be spaced as per Figure R307.1.

R307.2 Bathtub and shower spaces. Bathtub and shower floors and walls above bathtubs with installed shower heads and in shower compartments shall be finished with a nonabsorbent surface. Such wall surfaces shall extend to a height of not less than 6 feet (1829 mm) above the floor.

Q: Can a window be installed within a shower compartment?

A: The building official would need to determine if the window installation would be considered a nonabsorbent surface to a height of not less than 6 feet above the floor. This is also a safety glazing concern.

SECTION R308 GLAZING

R308.1 Identification. Except as indicated in Section R308.1.1 each pane of glazing installed in hazardous locations as defined in Section R308.4 shall be provided with a manufacturer's designation specifying who applied the designation, designating the type of glass and the safety glazing standard with which it complies, which is visible in the final installation. The designation shall be acid etched, sandblasted, ceramic-fired, laser etched, embossed, or be of a type which once applied cannot be removed without being destroyed. A label shall be permitted in lieu of the manufacturer's designation.

Exceptions:

1. For other than tempered glass, manufacturer's designations are not required provided the building official approves the use of a certificate, affidavit or other evidence confirming compliance with this code.
2. Tempered spandrel glass is permitted to be identified by the manufacturer with a removable paper designation.

Q: We have a situation where a hot tub is installed in the corner of the bathroom. There is a sitting area on two sides of the hot tub that are level with the top of the hot tub. We have a few questions regarding the glazing. If the lowest edge of the glazing of the two large windows in the corner are less than 60 inches above the standing surface of the hot tub, are they required to be safety glazing?

A: Yes. This area would be considered hazardous due to the possibility of slipping on the wet surface of the hot tub. See Figure 3-15.

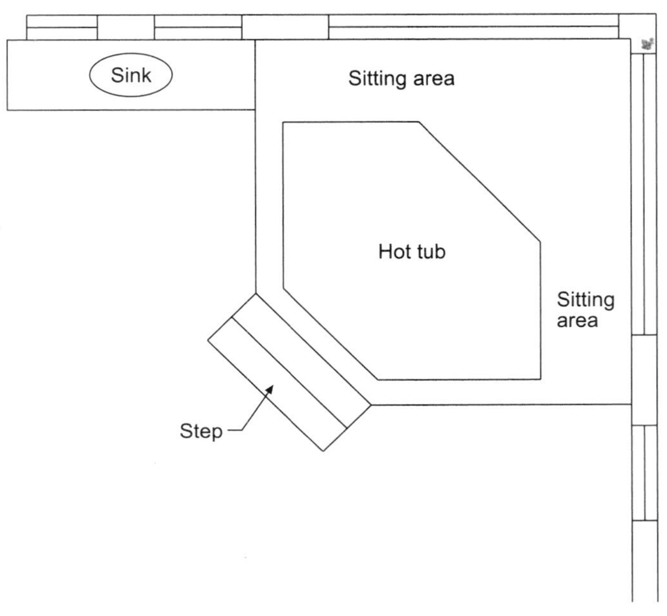

SAFETY GLAZING AT HOT TUB
FIGURE 3-15

Q: If the glazing was greater than 60 inches above the standing surface of the hot tub, but less than 60 inches above the sitting area, would the glazing area still need to be safety glazing?

A: No. The provision only addresses the level of the standing surface of the hot tub.

Q: We have one last question regarding that corner hot tub. There is a sink on the adjacent wall that has two individual windows directly above the sink. Are these windows required to be safety glazing?

A: No. The requirement for hazardous locations, or safety glazing, is based upon the glazing that is in the walls that enclose this hot tub. The enclosure stops at the edge of the hot tub.

Q: Does Item 6 in Section R308.4 apply to both of the casement windows A and D in the angled portion of the bay window in Figure 3-16? The drawing indicates a sliding patio door with a casement window on each side of the patio door. Only Panel B of the patio door slides. Panel C is in a fixed position.

A: Item 6 refers to glazing within a 24-inch arc of a door in a closed position. It is a widely held opinion that the intent of the requirement is based upon a door that opens for passage; therefore, the appropriate measurement would be taken from the nearest edge of the active Panel B. Due to its spatial separation of more than 24 inches, casement window D adjacent to this fixed panel is not considered to be in a hazardous location and thus would not require safety glazing.

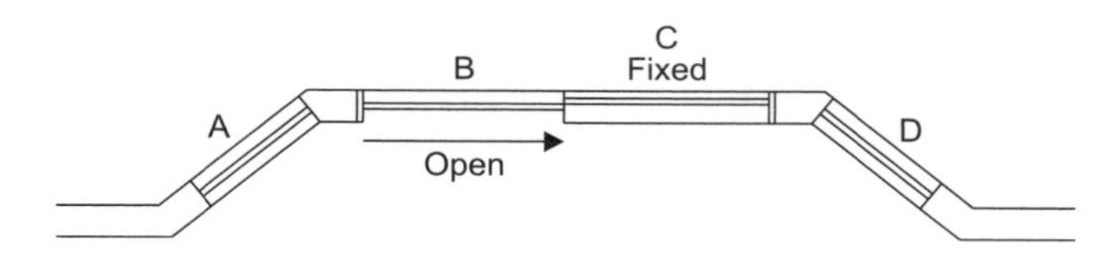

SAFETY GLAZING AT BAY WINDOW
FIGURE 3-16

R308.6.8 Curbs for skylights. All unit skylights installed in a roof with a pitch flatter than three units vertical in 12 units horizontal (25-percent slope) shall be mounted on a curb extending at least 4 inches (102 mm) above the plane of the roof unless otherwise specified in the manufacturer's installation instructions.

Q: Why is the skylight required to be installed on a curb extending at least 4 inches above the plane of the roof?

A: It is required for roofs with a slope less than three units vertical in 12 units horizontal (3:12 pitch) to allow a proper means of flashing the skylight to facilitate drainage past the skylight.

SECTION R309
GARAGES AND CARPORTS

R309.1 Opening protection. Openings from a private garage directly into a room used for sleeping purposes shall not be permitted. Other openings between the garage and residence shall be equipped with solid wood doors not less than $1^3/_8$ inches (35 mm) in thickness, solid or honeycomb core steel doors not less than $1^3/_8$ inches (35 mm) thick, or 20-minute fire-rated doors.

R309.1.1 Duct penetration. Ducts in the garage and ducts penetrating the walls or ceilings separating the dwelling from the garage shall be constructed of a minimum No. 26 gage (0.48 mm) sheet steel or other approved material and shall have no openings into the garage.

R309.1.2 Other penetrations. Penetrations through the separation required in Section R309.2 shall be protected by filling the opening around the penetrating item with approved material to resist the free passage of flame and products of combustion.

R309.2 Separation required. The garage shall be separated from the residence and its attic area by not less than $^1/_2$-inch (12.7 mm) gypsum board applied to the garage side. Garages beneath habitable rooms shall be separated from all habitable rooms above by not less than $^5/_8$-inch (15.9 mm) Type X gypsum board or equivalent. Where the separation is a floor-ceiling assembly, the structure supporting the separation shall also be protected by not less than $^1/_2$-inch (12.7 mm) gypsum board or equivalent. Garages located less than 3 feet (914 mm) from a dwelling unit on the same lot shall be protected with not less than $^1/_2$-inch (12.7 mm) gypsum board applied to the interior side of exterior walls that are within this area. Openings in these walls shall be regulated by Section R309.1. This provision does not apply to garage walls that are perpendicular to the adjacent dwelling unit wall.

Q: What fire-resistant separation is required between a residential dwelling and the attached garage?

A: "The garage shall be separated from the residence and its attic area by not less than ½-inch gypsum board applied to the garage side," per Section R309.2.

Q: Does the code refer to a minimum or maximum joint space between the sheets of gypsum board?

A: No. Reference should be made to the manufacturer of the gypsum board for any recommendations.

Q: Are the joints between the sheets of the ½-inch gypsum board required to be treated (joints filled/taped).

A: No.

Q: How is this ½-inch gypsum fastened to the framing members?

A: Refer to Table R702.3.5, "Minimum Thickness and Application of Gypsum Board." The answer is based on the type of fastener, spacing of fasteners, spacing of framing members, orientation of gypsum board to framing and thickness of gypsum board. For example, if the installer intended to use screws to install the ½-inch gypsum board vertically on a wall with studs spaced 16 inches on center, those screws would need to be Type S or W, be sufficiently long enough to penetrate the wood framing not less than $^5/_8$ inch and be spaced no more than 16 inches on center along framing members. Type S screws can be used for wood or sheet metal. Type W screws are for wood framing.

Q: We have a series of questions regarding Figure 3-17. There will be habitable space above the attached garage that will be accessible from the second story of the dwelling. The fire-resistant separation is going to be applied to the house/garage common wall, and it extends on the ceiling of the garage due to the tuck-under situation. What fire-rated material does the code require on the ceiling of the attached garage?

A: Garages shall be separated from all habitable rooms above by not less than $^5/_8$-inch Type X gypsum board or equivalent.

GARAGE BELOW HABITABLE SPACE
FIGURE 3-17

Q: How is this $^5/_8$-inch Type X gypsum board fastened to the garage ceiling?

A: Note e of Table R702.3.5 states, "Type X gypsum board for garage ceilings beneath habitable rooms shall be installed perpendicular to the ceiling framing and shall be fastened at maximum 6 inches on center by minimum 1 $^7/_8$ inch 6d coated nails or equivalent drywall screws."

Q: Can staples be substituted for the nails or screws noted in the previous question for the garage ceiling?

A: The footnote makes no reference to the use of staples for this garage ceiling. Although this is not a listed assembly, the purpose is for fire resistance. The building official could allow the use of staples if testing indicated the staples provided the same holding power as the required nails under fire conditions.

Q: The 24-inch front roof overhang of the garage butts into the exterior house wall. Does the ½-inch gypsum board of the garage separation need to extend into the soffit area?

A: It depends on how the complete separation as required by the code is achieved. In this case, the garage has a $^5/_8$-inch Type X gypsum board ceiling as part of the separation. The front and back walls supporting the ceiling will require ½-inch gypsum board. These elements in addition to the common wall provide a complete separation and the soffit would not require further protection. On the other hand, if the garage space was open to the soffit area, then the ½-inch gypsum board on the common wall would need to extend into the soffit area out to the fascia where this front overhang terminates into the second floor wall to provide a complete separation from the dwelling.

Q: Since installation of $^5/_8$-inch Type X gypsum board is mandated on the ceiling of the garage, are the garage walls required to be covered with ½-inch gypsum board also?

A: The garage walls supporting the floor/ceiling assembly require ½-inch gypsum board per Section R309.2.

Q: Would the exterior stud wall below the gable end of the attached garage be considered part of the "structure supporting the separation" for the purposes of determining the location of the ½-inch gypsum board?

A: No. The ceiling separation in this case is provided by the floor/ceiling assembly, meaning the entire structure of the floor, including the joists or trusses. Unless the gable end wall provides structural support to the floor system above, it does not require application of ½-inch gypsum board.

Q: The code requires the habitable space above the garage be protected with one layer of $^5/_8$-inch Type X gypsum board on the garage ceiling. Could wood structural panel sheathing be used to protect the supporting walls below this ceiling?

A: The code requires the structure supporting the separation to be protected by not less than ½-inch gypsum board or equivalent. If wood structural panel sheathing was proposed as the equivalent, the designer/builder would need to provide supporting documentation to verify the fire-resistant properties of the wood structural panel based upon the thickness and other properties. Table 720.6.2(1) of the IBC lists the fire properties of tested wood structural panels in minutes of fire resistance. The APA-Engineered Wood Association has test data of thermal testing of wood structural panels based on ASTM E 119. Both of these references may be useful in making this determination.

Q: Does the code require the installation of a door between the dwelling and its attached garage?

A: No. The code requires proper separation, but does not specifically require a door to be installed that would allow direct access between the dwelling and the attached garage. Access to the garage could be obtained through the overhead door only.

Q: If a door is installed between the dwelling and garage, is the door required to be fire-rated?

A: No. Section R309.1 requires "openings between the garage and residence shall be equipped with solid wood doors not less than $1^3/_8$ inch in thickness, solid or honeycomb core steel doors not less than $1^3/_8$ inch thick, or 20 minute fire-rated doors." The wood door and honeycomb steel door are not required to be fire rated. The "20-minute fire-rated door" is another option. See Figure 3-18.

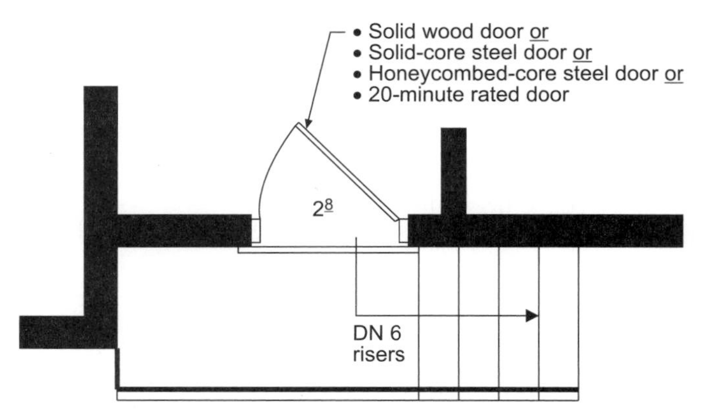

DWELLING ACCESS DOOR FROM ATTACHED GARAGE
FIGURE 3-18

Q: Is a self-closing device required for the door between the residence (dwelling) and the attached garage?

A: No. There is no requirement for either a self-closing device or a latching device.

Q: Is the frame for the door between the dwelling and the attached garage required to be steel, or part of a rated assembly?

A: No. The frame can be a simple wood frame.

R311.5.4 Landings for stairways. There shall be a floor or landing at the top and bottom of each stairway.

> **Exception:** A floor or landing is not required at the top of an interior flight of stairs, including stairs in an enclosed garage, provided a door does not swing over the stairs.

Q: Is a landing required for the garage side of the door between the dwelling and the garage?

A: It depends on the swing of the door. Refer to Section R311.5.4: "A floor or landing is not required at the top of an interior flight of stairs, including stairs in the enclosed garage, provided a door does not swing over the stairs."

Q: Can a 20-minute fire-resistance-rated door between the dwelling and garage contain glazing?

A: Yes, if approved by the building official. This is not a fire door assembly, but a building official might reasonably conclude that fire-resistant safety glazing would provide some equivalent protection. NFPA 80, Section 1-7, "Glazing Material in Fire Doors," states that only labeled fire- resistance-rated glazing material meeting applicable safety standards shall be used in fire door assemblies. Testing determines the maximum size of glazing material permitted in 20-minute-rated fire doors.

Q: Does the code require the door between the dwelling and the garage to swing in any particular direction?

A: No. The door can swing into the dwelling or into the garage. If the door swings into the garage, it will also need a proper landing.

Q: Our staff is looking at a proposal for a detached garage with a bonus room above, and we have a few related questions regarding the plans. The plans are for the bonus room to be used for living space, a library and recreation space. The only exit/entrance to this bonus room is a code-complying stairway through the garage space below. Does this situation meet the code requirement for an exit?

A: No. If this space is considered part of the dwelling, then it will also require an exit (Section R311.4.1). The required exit door shall provide for direct access from the habitable portions of the dwelling to the exterior without requiring travel through a garage. The stair from the garage to this bonus room would not comply with code as a required exit.

Q: Regarding this bonus room above the garage, an elevated enclosed walkway is proposed as an option that would be constructed over to the second floor of the dwelling 10 feet away from this garage. Would the proposed stairway between the garage and the bonus room that was originally proposed now comply with code because the occupants of the bonus room can exit through the front door?

A: Yes, but the requirements for fire separation between the habitable space and the garage need to be met.

Q: We have a plan for a two-story dwelling that extends over a tuck-under garage at one end. This structure is 6 feet from the lot line. The interior walls will be protected with ½-inch gypsum board and the ceiling will be protected with ⅝-inch Type X gypsum. Do the exterior sides of the garage wall need to be protected with ½-inch gypsum board because they are supporting the horizontal fire separation between the house and the garage?

A: No. These exterior walls of the garage are only protected on the inside face. They are not constructed as 1-hour-rated wall assemblies.

Q: In this situation above, are there any special requirements for windows in the garage walls to be fire-rated glass assemblies?

A: No. There are no requirements for protection of openings in exterior walls in this structure that is 6 feet from the lot line.

Q: We regularly see permit applications for dwellings with rear walkouts and large atrium/patio doors that might allow the occupants to drive in and store lawn mowers and other gasoline-powered vehicles in the basement. Should we require fire separation between this space and the balance of the dwelling?

A: The plan should indicate the use or occupancy of each room in the structure. If the space is designated as a garage, it should be separated per Section R309.2. See Figure 3-19.

Q: Could a building department restrict the size of a walkout level patio door to prohibit the entrance of riding lawn mowers and other vehicles into the space?

A: No.

Q: Section R309.1.2, "Other penetrations," states that penetrations "shall be protected by filling the opening around the penetrating item with approved material to resist the free passage of flame and products of combustion." What kinds of products does this refer to, and what does "approved" refer to?

A: This fire-resistant separation is not a listed assembly. There are no specific items listed in the code for the purpose of sealing these penetrations. The section implies that some material, such as a mineral wool or fiberglass, cement-based products, a fire-stopping caulk or other material approved by the building official be used to fill the opening around the penetration.

DOOR OPENINGS AT REAR WALKOUT OF DWELLING
FIGURE 3-19

Q: Are there any restrictions on the use of plastic pipe, such as PVC or ABS, through the ½-inch gypsum board separating the garage and the dwelling?

A: No. It is permitted. See Figure 3-20.

PLASTIC PIPE PENETRATION OF GARAGE AND DWELLING SEPARATION
FIGURE 3-20

Q: If an attic access, such as a pull-down ladder, is installed in the fire-resistant ceiling of the garage, does it need to meet the same requirements as the service door into the dwelling?

A: Although the code does not make a specific reference to this attic access (or scuttle), it would still need to maintain the integrity of the fire separation required between the garage and the dwelling.

Q: A dwelling is constructed with an attached garage. Inside the attached garage there is a forced air furnace that provides heat for the dwelling through metal ducts. Is this permitted by code?

A: Yes. In Sections R309.1.1 and R309.1.2, the code addresses the issue of ducts in the garage and other penetrations. The sections do not address the use of a fire damper, but it may be considered an acceptable alternative for having duct openings in the garage. The final approval is still subject to the local building official.

Q: Would this forced-air furnace in the attached garage require any fire-resistance-rated separation from the parking area of this residential garage?

A: No. Although, some type of impact protection from the vehicle is required per the mechanical provisions of the IRC in Section M1307.3.1.

Q: A one-story dwelling has an attached garage at one end. The attic area above the dwelling and garage is not habitable space. Instead of installing the fire separation at the house/garage common wall up to the roof sheathing, the builder has chosen to only install gypsum board on the garage ceiling and all four walls. The attic area between the dwelling and attached garage is open. Does the gypsum board on the garage ceiling need to be ⁵⁄₈-inch Type X, or ½-inch regular gypsum board?

A: It needs to be at least ½ inch regular gypsum board.

Q: To follow-up on the previous question, if the fire separation was provided on the ceiling of the garage and on the supporting walls inside the garage, would draftstopping be required in the attic between the garage and the dwelling?

A: No. The attic above the dwelling can be open to the attic above the garage.

Q: The attic above a detached garage is designed with a 7-foot ceiling and trusses capable of supporting some storage. The owner wants to use a pull-down ladder for access to this space. Would this now require a code-complying stairway because of the ceiling height in this attic storage area?

A: No. There are no provisions in the code that would prevent the use of a ladder from being used for access to this attic storage area.

Q: In the situation above, if a stairway was installed to this attic storage area, would the stair riser and run need to comply with code?

A: Yes. If it looks like a stairway, then it needs to be constructed to meet the minimum requirements for a stairway. This would include proper rise, run, stair width, handrail, guards, etc.

Q: Does the attached garage require any mechanical ventilation to exhaust the fumes from vehicles?

A: No. There are no provisions in the code that would require any exhaust ventilation in the attached garage to remove the exhaust fumes of the vehicle.

R309.3 Floor surface. Garage floor surfaces shall be of approved noncombustible material.

The area of floor used for parking of automobiles or other vehicles shall be sloped to facilitate the movement of liquids to a drain or toward the main vehicle entry doorway.

Q: There has been a lot of discussion in our office about what type of garage floor material could be considered "approved noncombustible" according to the code. We have also seen applications such as plastic panels and epoxy coatings applied onto a concrete floor after the curing has taken place. Could you offer some assistance?

A: The word "approved" refers to approved by the building official. In Section R202, "Definitions," a "Noncombustible material" is defined as a material that passes the test procedure for defining noncombustibility of elementary materials set forth in ASTM E 136. The product specifications should indicate any testing conducted on those products.

Q: Could hard-packed gravel or crushed rock serve as the garage floor?

A: No. Although gravel or crushed rock are noncombustible, the section also requires the garage floor surface to be sloped to facilitate the movement of liquids to a drain or toward the main vehicle entry doorway. The objective is to use a material for the garage floor that will not contribute to a fire in the garage, but one that will keep the liquid that could drop off a vehicle from entering into the soil within the garage perimeter.

Q: Are smoke alarms or carbon monoxide alarms required in an attached garage?

A: No. The requirements for smoke alarms are contained in Section R313. There are no requirements for carbon monoxide alarms in the IRC.

Q: Does the code require the floor of an attached garage to be lower than the floor of the dwelling that it is attached to?

A: The code does not require such separation of elevation. The garage floor and the interior floor level of the dwelling can be at the same elevation. See Figure 3-21.

ATTACHED GARAGE FLOOR AT SAME ELEVATION AS DWELLING FLOOR
FIGURE 3-21

Q: Does the code require any sort of door gasket on the bottom of the overhead garage door to prevent water from entering?

A: No.

SECTION R310 EMERGENCY ESCAPE AND RESCUE OPENINGS

R310.1 Emergency escape and rescue required. Basements and every sleeping room shall have at least one operable emergency escape and rescue opening. Such opening shall open directly into a public street, public alley, yard or court. Where basements contain one or more sleeping rooms, emergency egress and rescue openings shall be required in each sleeping room, but shall not be required in adjoining areas of the basement. Where emergency escape and rescue openings are provided they shall have a sill height of not more than 44 inches (1118 mm) above the floor. Where a door opening having a threshold below the adjacent ground elevation serves as an emergency escape and rescue opening and is provided with a bulkhead enclosure, the bulkhead enclosure shall comply with Section R310.3. The net clear opening dimensions required by this section shall be obtained by the normal operation of the emergency escape and rescue opening from the inside. Emergency escape and rescue openings with a finished sill height below the adjacent ground elevation shall be provided with a window well in accordance with Section R310.2. Emergency escape and rescue openings shall open directly into a public way, or to a yard or court that opens to a public way.

> **Exception:** Basements used only to house mechanical equipment and not exceeding total floor area of 200 square feet (18.58 m^2).

R310.1.1 Minimum opening area. All emergency escape and rescue openings shall have a minimum net clear opening of 5.7 square feet (0.530 m^2).

Exception: Grade floor openings shall have a minimum net clear opening of 5 square feet (0.465 m^2).

R310.1.2 Minimum opening height. The minimum net clear opening height shall be 24 inches (610 mm).

R310.1.3 Minimum opening width. The minimum net clear opening width shall be 20 inches (508 mm).

R310.1.4 Operational constraints. Emergency escape and rescue openings shall be operational from the inside of the room without the use of keys, tools or special knowledge.

Q: What is the intent of the emergency escape and rescue window?

A: It is a means of escape for the occupants directly to the exterior of the dwelling, and a means for access by emergency personnel.

Q: Does the code allow a door to be used as the emergency escape and rescue opening in a situation where there is not a complying window?

A: Yes. The door needs to meet the same requirements in Section R310.1.

Q: Does a basement without habitable space require an emergency escape and rescue opening?

A: Yes. Previous editions of the IRC required emergency escape and rescue openings in basements with habitable space, but the words "habitable space" have been removed. Emergency escape and rescue openings are now required in all basements except those with 200 square feet or less of floor area housing only mechanical equipment.

Q: Is an emergency escape and rescue opening required in a basement if the space is not finished?

A: Yes. The emergency escape and rescue opening is required whether the basement is finished or not finished.

Q: At what minimum ceiling height in a basement would an emergency escape and rescue opening be required?

A: The definition of "Basement" does not address a ceiling height, and a determination regarding escape and rescue openings cannot be based on ceiling height alone. All basements greater than 200 square feet require an emergency escape and rescue opening. If there is doubt as to whether a space is a basement, the building official must make a determination based on available information. For example, criteria may include the use or usability of the space and the means of access to the space.

Q: As a plans examiner, sometimes plans are submitted for dwellings that indicate rooms with closets, and floor areas that appear to indicate a room that might be used as a sleeping room. How do I determine if an emergency escape and rescue opening is required?

A: An emergency escape and rescue opening is required in all sleeping rooms. The designer/drafter should indicate on the plan how each room is to be used. A closet shown on the plan in a room does not necessarily mean that it is to be used as a sleeping room. The code does not require a closet in any room.

Q: The owner of a dwelling would like to create a sleeping room in an existing basement. Currently, that space is being used for storage and there are no windows. The adjoining room is a recreational space that contains three 24-inch by 36-inch casement windows. The owner does not want to install any additional windows, and has asked if he or she could take down the existing wall between the two rooms and create one very large sleeping room containing three 24-inch by 36-inch casements. Would this situation comply with code? See Figure 3-22.

A: Yes, if at least one of the existing 24-inch by 36-inch casement windows meets the minimum requirements of Section R310.1 for emergency escape and rescue openings and the space complies with the other applicable requirements of the code, such as light, ventilation, headroom and smoke alarm.

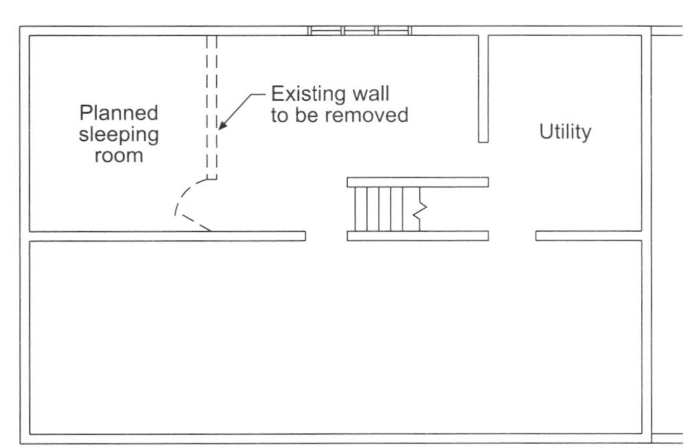

SLEEPING ROOM IN BASEMENT LEVEL OF DWELLING
FIGURE 3-22

Q: Our inspection staff has a difference of opinion on where the 44 inches is measured from the floor. Although the code refers to the sill of the window, some have suggested that the intent is to measure from the lowest portion of the actual clear opening of the window.

A: The 44 inches is measured to the top of the sill. In Figure 3-23 the sill is located where the girl's left foot is placed.

FATHER ASSISTING DAUGHTER THROUGH WINDOW
FIGURE 3-23

Q: We believe that the top of the sill is intended to imply the point at which the opening in the emergency escape and rescue window occurs. Is this true?

A: No. Although other provisions in Section R310 specify dimensions of the net clear opening, this provision is specific to the sill. It is assumed that some portion of the window bottom track or stop will be at an elevation slightly above the sill.

Q: A homeowner created a sleeping room in an existing basement that included a new emergency escape and rescue window. Unfortunately, he installed the window about 50 inches above the basement floor. He has asked if he could install one single step to try to meet the 44 inch maximum allowed by the code. Could a step be used to meet this 44-inch maximum dimension?

A: No. The measurement is taken from the floor level. See Figure 3-24.

Q: Could a 36-inch by 36-inch landing be considered a floor for this purpose?

A: No. The measurement is taken from the floor level. See Figure 3-25.

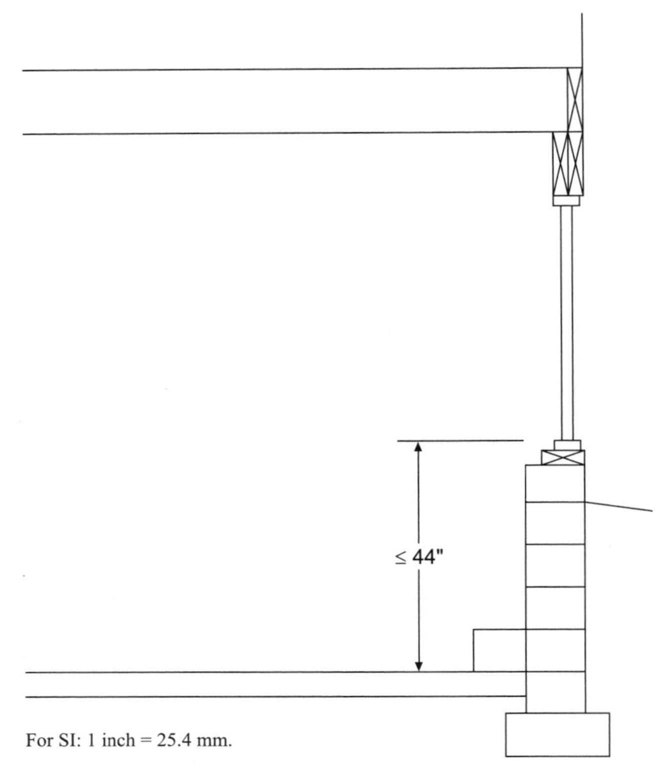

STEP AT EMERGENCY ESCAPE AND RESCUE WINDOW OPENINGS
FIGURE 3-24

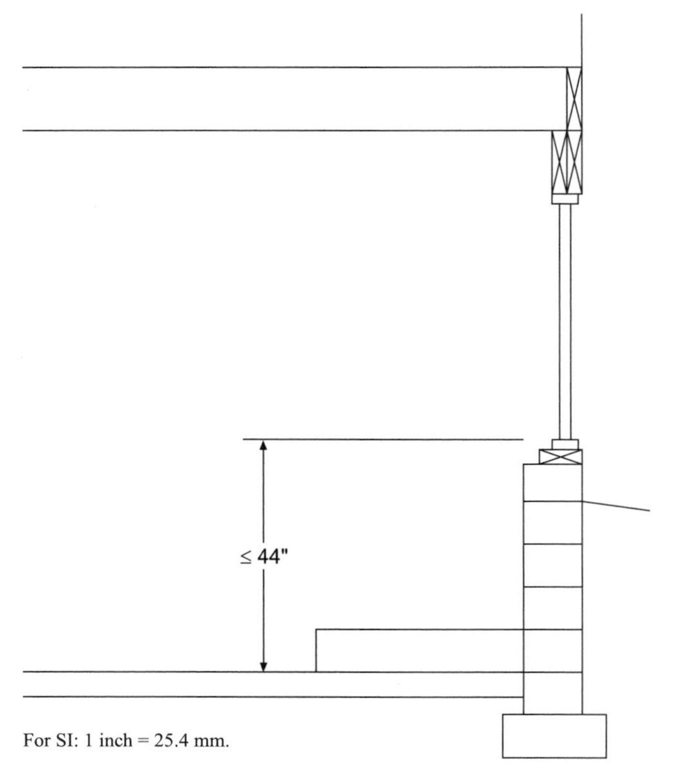

LANDING AT EMERGENCY ESCAPE AND RESCUE WINDOW OPENING
FIGURE 3-25

Q: A permit application was submitted for the construction of a new house. On the plan, all three of the rear sleeping rooms on the second floor were drawn indicating a sliding patio door as the emergency escape and rescue component. The permit applicant does not plan on constructing the rear deck along the rear of the house until a later date. Can a sliding patio door serve as the required emergency escape and rescue opening if no deck is in place?

A: Yes. A proper guard could be constructed on the outside of each of these sliding patio doors that would provide the fall protection, yet the occupant in the sleeping room could climb over the 36-inch-high guard as long as it still met the minimum net clear opening requirements for emergency escape and rescue opening.

Q: When measuring the minimum net clear opening for emergency escape and rescue opening when using a double-hung window, can the net clear area of the upper sash be included since sashes are normally not that difficult to remove?

A: No. In Section R310.1 it states, "the net clear opening dimension required by this section shall be obtained by the normal operation of the emergency escape and rescue opening…" The extra effort to lift and pull the sash is not considered to be normal operation. For children and physically challenged adults, it might prove to be too much in the event of an emergency.

Q: Will the installation of a 13R or 13D sprinkler system be allowed as a substitution for required emergency escape and rescue openings in basements and sleeping rooms?

A: No. Section R310 does not provide any trade-offs for the emergency escape and rescue openings.

Q: Can an emergency escape and rescue opening in a sleeping room open into the garage?

A: No. Section R310.1 specifically requires it to open directly into a public street, public alley, yard or court. In addition, Section R309.1 states, "openings from a private garage directly into a room used for sleeping purposes shall not be permitted."

R310.2 Window wells. The minimum horizontal area of the window well shall be 9 square feet (0.9 m^2), with a minimum horizontal projection and width of 36 inches (914 mm). The area of the window well shall allow the emergency escape and rescue opening to be fully opened.

Exception: The ladder or steps required by Section R310.2.1 shall be permitted to encroach a maximum of 6 inches (152 mm) into the required dimensions of the window well.

R310.2.1 Ladder and steps. Window wells with a vertical depth greater than 44 inches (1118 mm) shall be equipped with a permanently affixed ladder or steps usable with the window in the fully open position. Ladders or steps required by this section shall not be required to comply with Sections R311.5 and R311.6. Ladders or rungs shall have an inside width of at least 12 inches (305 mm), shall project at least 3 inches (76 mm) from the wall and shall be spaced not more than 18 inches (457 mm) on center vertically for the full height of the window well.

Q: Can the ladder encroach into any area of the window well?

A: Yes. It may project 6 inches into the minimum required dimensions of the window well from the sides or wall of the dwelling. See Figure 3-26.

LADDER IN WINDOW WELL
FIGURE 3-26

Q: Could a ladder be installed on two sides and project 6 inches from each of those two sides?

A: No. If the window well was 36 inches by 36 inches, it would reduce the actual area of escape down to 24 inches wide. That was not the intent of the code provision.

Q: In the situation of an emergency escape and rescue window in a basement that opens into a window well, are the minimum inside dimensions (36 inches by 36 inches) of the window well measured to the inside face of the window after it is cranked out in the open position?

A: No. The operation of the window has no effect on the required inside dimensions of the window well, provided the window opens fully. The 36-inch by 36-inch dimension is measured as the inside dimensions of the window well.

Q: Does the code have any requirements to address the construction of the ladder inside of the egress window well?

A: Although the code does address the width and spacing of the rungs, it does not contain any specific requirements related to the actual material or installation. It does imply that the ladder be constructed of materials approved for exterior use, be adequate to carry the weight of the person using it for emergency escape or rescue and be adequately fastened to the window well.

Q: Does the grade level inside of the window well need to meet the provisions in Section R319 for protection against decay?

A: Yes. The provisions in Section R319, "Protection Against Decay," are applicable. These generally require the sill plate to be naturally durable or preservative-treated wood if less than 8 inches above exposed ground. Wood siding, sheathing and framing must be at least 6 inches above the ground. Applying this section, a 6-inch clearance is required below elements of the wall or window that are not naturally durable or preservative-treated wood.

Q: How high does the ladder need to extend upward?

A: The ladder should at least reach the top of the window well. See Figure 3-27.

VIEW OF EMERGENCY ESCAPE AND RESCUE WINDOW OPENING FROM BASEMENT LEVEL
FIGURE 3-27

Q: How high does the window well need to extend above the adjacent grade level?

A: The code does not address a height above grade. It should extend above the grade level to prevent the soil from entering into the window well.

R310.4 Bars, grilles, covers and screens. Bars, grilles, covers, screens or similar devices are permitted to be placed over emergency escape and rescue openings, bulkhead enclosures, or window wells that serve such openings, provided the minimum net clear opening size complies with Sections R310.1.1 to R310.1.3, and such devices shall be releasable or removable from the inside without the use of a key, tool, special knowledge or force greater than that which is required for normal operation of the escape and rescue opening.

Q: During a final inspection of a new house, our inspector observed a cover on top of an emergency escape and rescue opening window well. It was constructed of some plastic material and was connected to the window well with wing nuts that were tightened from the inside. Section R310.4 states, "removable without the use of keys, tools, ... special knowledge." Our inspector did not approve the installation.

A: Wing nuts do not comply with this requirement. This code provision is intended to allow the installation of covers over emergency escape and rescue opening window wells that would help keep rain and snow from entering into the window well itself, and still allow the emergency escape of the occupant. The use of wing nuts to hold down this cover would certainly hinder the emergency escape of a child that is unfamiliar with the method of removal of the wing nuts (special knowledge), or of anybody in a panic situation where the inside of the window well is starting to fill with smoke in the event of a fire.

R310.5 Emergency escape windows under decks and porches. Emergency escape windows are allowed to be installed under decks and porches provided the location of the deck allows the emergency escape window to be fully opened and provides a path not less than 36 inches (914 mm) in height to a yard or court.

Q: We have a situation where a deck completely spans and rests on the window well concrete foundation. Does the 36 inches minimum under a deck imply in all directions or at least in one direction?

A: The code only requires one path not less than 36 inches in height to a yard or court. See Figure 3-28.

EMERGENCY ESCAPE UNDER DECK
FIGURE 3-28

Q: What is the minimum width of this path?

A: Although not specifically noted in this section, all other references in the code would imply that the path would be at least 36 inches wide.

SECTION R311 MEANS OF EGRESS

R311.1 General. Stairways, ramps, exterior egress balconies, hallways and doors shall comply with this section.

Q: Where does the means of egress system end?

A: The IRC provides for a means of egress to get the occupants safely out of the building to grade. Generally speaking, it ends where the occupant walks out the front door, down the steps and onto the landing at grade level. Exit discharge is not regulated by the IRC beyond that point. Although the IBC contains some specific language about the exit discharge in the means of egress system extending to a public way, the IRC does not contain any such language.

Q: A townhouse project was constructed recently in our municipality (see Figure 3-29). The main entrance doors of the end units open to a sidewalk that turns 90 degrees and continues to a set of concrete stairs down the embankment. Does the concrete stair on the embankment need to meet the code requirements for rise, run, width of stair tread and any handrail requirements?

A: No. The concrete stairway on the embankment is not regulated by the code.

CONCRETE STAIRWAY ON EMBANKMENT
FIGURE 3-29

R311.2 Construction.

R311.2.1 Attachment. Required exterior egress balconies, exterior exit stairways and similar means of egress components shall be positively anchored to the primary structure to resist both vertical and lateral forces. Such attachment shall not be accomplished by use of toenails or nails subject to withdrawal.

Q: Does the provision for attachment in Section R311.2.1 apply to all exterior exit landings and stairs?

A: Yes. Such landings and stairs at any exterior door are considered components of the means of egress system. Positive attachment to the structure is essential for safety at these exterior locations. See Figure 3-30.

ATTACHMENT OF EXTERIOR LANDING TO DWELLING
FIGURE 3-30

Q: Does this section imply the use of lag screws or carriage bolts to anchor this ledger board?

A: Although this section does not allow the use of toe nails or nails subject to withdrawal, it does not specify any other type of fastener. This could include the use of lag screws, bolts or approved manufactured metal connectors.

Q: The prescriptive provisions in the code do not appear to address the minimum type, size or quantity of fasteners for the ledger board. How does a designer or builder determine the correct size of anchors, amount of anchors and if they are installed in the correct location?

A: The designer needs to evaluate the loading on the ledger or rim joist and prescribe an anchoring system based on the strength characteristics of the particular anchor. This may include the use of lag screws or bolts. The location and spacing of the particular anchoring device, along with the diameter and embedment should be determined and specified as part of the design of the structure.

Q: If this ledger is treated wood, do the fasteners (screws or bolts) need to be galvanized?

A: Yes, or other materials approved for the use. Refer to the Q&A for Section R319, "Protection Against Decay."

R311.2.2 Under stair protection. Enclosed accessible space under stairs shall have walls, under stair surface and any soffits protected on the enclosed side with 1/2-inch (13 mm) gypsum board.

Q: The basement stairs of all of our homes have walls on both sides with drywall applied to the stairway side. Since the drywall continues down to the basement floor, a small area is created below the stairs and is often used for storage. Access to this under-stair area is typically an opening approximately 3 feet wide by 7 feet high. We have several municipalities requiring us to apply drywall to the underside of the stairs, even though there is no door. We would appreciate some clarification of what "enclosed" means.

A: "Enclosed" means to be surrounded on all sides by walls with entry to the under-stair area through a door or access panel. Since the "enclosure" you describe is open on the end or partially open on one side, it is not considered enclosed and protection with 1/2-inch gypsum board on the underside of the stairs is not required by the code. This may be a case where best practices would suggest installing the drywall at the time of construction but such is not mandated. If a door is installed, then the ceiling in this area would require the application of 1/2-inch gypsum board. The intent of this section is to provide minimal fire protection for the stairway, which is part of the means of egress. Enclosed accessible spaces beneath stairways are typically used for storage. A high combustible fuel load out of sight poses a significant fire risk to the building occupants such that they may not be able to quickly detect the fire and evacuate the area.

Q: Does the code require a 1-hour fire-resistant barrier below enclosed stairs?

A: No. It only requires 1/2-inch gypsum board applied to walls, ceilings and soffits on the enclosed side under stairs. See Figure 3-31.

UNDER-STAIR PROTECTION
FIGURE 3-31

Q: A contractor has placed a furnace below an enclosed stair prior to the installation of any 1/2-inch drywall. Some of the warm air supply ductwork is fastened directly to the underside of the wood floor joists of the stair landing above. Does this violate the intent of the code?

A: The installation as described above does not comply with the code. The 1/2-inch gypsum board needs to be installed on the bottom side of the wood floor joists and the ducts can then go below the gypsum board.

R311.3 Hallways. The minimum width of a hallway shall be not less than 3 feet (914 mm).

Q: We understand the minimum width of a hallway is 36 inches. Where is this dimension measured? We have different opinions about if it is measured to the trim, face of the wall finish or rough framing.

A: The 36-inch minimum dimension is measured from the face of the finish material (such as the gypsum board). The code does not address any projections from the side for trim as it does for stairways.

R311.4 Doors.

R311.4.1 Exit door required. Not less than one exit door conforming to this section shall be provided for each dwelling unit. The required exit door shall provide for direct access from the habitable portions of the dwelling to the exterior without requiring travel through a garage. Access to habitable levels not having an exit in accordance with this section shall be by a ramp in accordance with Section R311.6 or a stairway in accordance with Section R311.5.

Q: Does the front entry door of a dwelling need to comply with Section R311.4.1 as the main exit door? Could the door located at the rear of the dwelling be considered as the required egress door?

A: Section R311.4.1 requires that each dwelling unit be provided with not less than one exit door complying with Section R311.4. This required exit door is not mandated to be the front door of the dwelling unit. The door located at the rear of the dwelling could be considered the main exit door if it met the requirements in Section R311.4.2.

R311.4.2 Door type and size. The required exit door shall be a side-hinged door not less than 3 feet (914 mm) in width and 6 feet 8 inches (2032 mm) in height. Other doors shall not be required to comply with these minimum dimensions.

Q: If the rear door is utilized as the required egress door, could this door be a sliding patio door?

A: No. The required exit door is required to be a side-hinged door as noted in Section R311.4.2.

Q: One of our homes has a side-hinged swinging door at the rear of the dwelling that was designated as the required exit door. The door is located more than 30 inches above the grade, but the customer does not want to install a deck at this time. We have proposed to the building official the installation of a guard across this door opening until the deck is constructed. The building official informed us that this door cannot serve as the required exit door because of the guard. Is this true?

A: Yes. This required exit door must provide unobstructed egress.

Q: We have a situation where a single-family dwelling has an attached maid's quarters. The entrance is separate from the primary entrance to the dwelling. Should the maid's entrance door comply with the minimum size and type of door noted in Section R311.4.2?

A: If this area is separate from the primary dwelling (that is, if this building meets the definition of a "Two-family dwelling"), each dwelling unit will require at least one complying egress door. The separation will also need to meet the fire-resistance requirements of Section R317.1

Q: If this maid's quarters had a common interior door to the primary residence, would the separate complying entrance door still be required?

A: Not necessarily. If the determination is that this building is a single dwelling unit, only one complying exit door would be required. The requirements in the local zoning code should be checked for compliance related to the living arrangement.

Q: Is there a minimum size requirement for doors that open into sleeping rooms?

A: No. There is no size requirement for doors other than the one required exterior exit door.

R311.4.3 Landings at doors. There shall be a floor or landing on each side of each exterior door. The floor or landing at the exterior door shall not be more than 1.5 inches (38 mm) lower than the top of the threshold. The landing shall be permitted to have a slope not to exceed 0.25 unit vertical in 12 units horizontal (2-percent).

Exceptions:

1. Where a stairway of two or fewer risers is located on the exterior side of a door, other than the required exit door, a landing is not required for the exterior side of the door provided the door, other than an exterior storm or screen door does not swing over the stairway.
2. The exterior landing at an exterior doorway shall not be more than $7^3/_4$ inches (196 mm) below the top of the threshold, provided the door, other than an exterior storm or screen door does not swing over the landing.
3. The height of floors at exterior doors other than the exit door required by Section R311.4.1 shall not be more than $7^3/_4$ inches (186 mm) lower than the top of the threshold.

The width of each landing shall not be less than the door served. Every landing shall have a minimum dimension of 36 inches (914 mm) measured in the direction of travel.

Q: The exceptions seem very similar, and our staff would appreciate your assistance in helping us understand these revised exceptions. We have a series of scenarios. Figure 3-32 shows the main exit door required by Section R311.4.1. The landing is 1½ inches below the top of the threshold. We believe that this main exit door can swing inward, or over the exterior landing. Is this correct?

A: Yes.

Q: In Figure 3-32 we believe the answer does not change with the addition of a storm or screen door. Is this correct?

A: Yes.

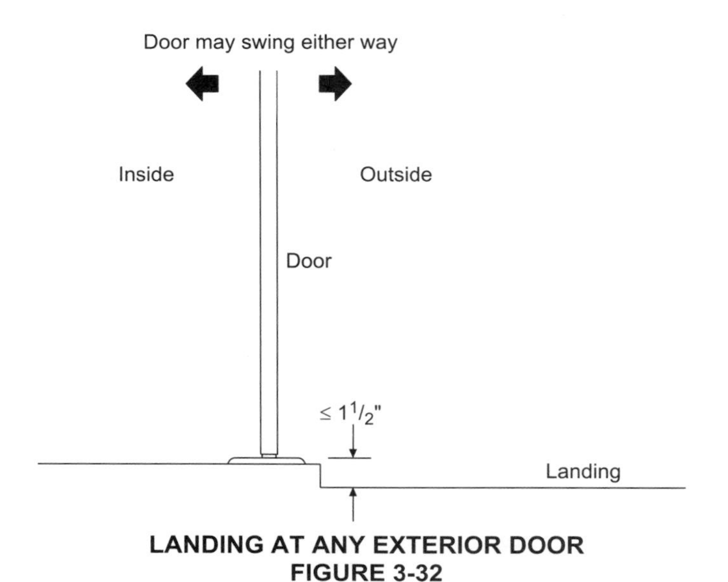

LANDING AT ANY EXTERIOR DOOR
FIGURE 3-32

Q: Figure 3-33 shows the main exit door required by Section R311.4.1. The landing is 7¾ inches below the top of the threshold. We believe that this main exit door cannot swing outward over the exterior landing. Is this correct?

A: Yes.

Q: In Figure 3-33 we believe that a storm door or screen can swing over the exterior landing as noted in Exception 2. Is this correct?

A: Yes.

Q: Figure 3-34 shows an exterior door with a two-riser stair down to grade level. We believe that this situation, noted as Exception 1, can only be used for exterior doors that are not the designated exit door required by Section R311.4.1. Is this correct?

A: Yes. All other exterior doors can use this provision.

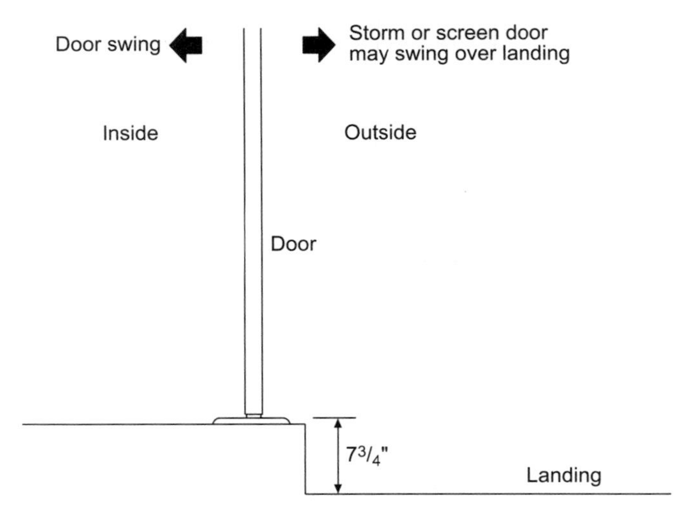

DOOR SWING AT EXTERIOR DOOR
FIGURE 3-33

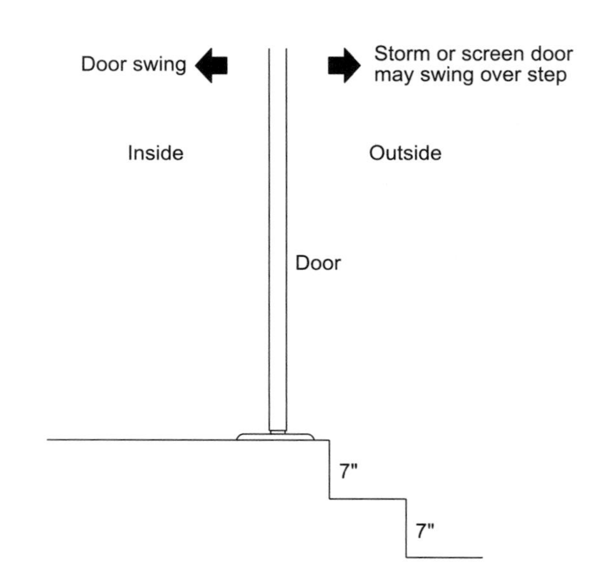

TWO-RISER STAIR AT EXTERIOR DOOR OTHER THAN REQUIRED EXIT
FIGURE 3-34

R311.4.4 Type of lock or latch. All egress doors shall be readily openable from the side from which egress is to be made without the use of a key or special knowledge or effort.

Q: Does this section prohibit the use of double-keyed deadbolt locks?

A: Yes.

R311.5 Stairways.

R311.5.1 Width. Stairways shall not be less than 36 inches (914 mm) in clear width at all points above the permitted handrail height and below the required headroom height.

Handrails shall not project more than 4.5 inches (114 mm) on either side of the stairway and the minimum clear width of the stairway at and below the handrail height, including treads and landings, shall not be less than 31.5 inches (787 mm) where a handrail is installed on one side and 27 inches (698 mm) where handrails are provided on both sides.

Exception: The width of spiral stairways shall be in accordance with Section R311.5.8.

Q: The code appears to require at least one code-complying egress stair in a dwelling. Could a secondary stair, such as a deck stair, or possibly a second stair in the garage to an upper level storage area be constructed that does not comply with the rise or run, stair width or headroom?

A: No. All stairs regulated by the code need to comply with the code requirements.

R311.5.3 Stair treads and risers.

R311.5.3.1 Riser height. The maximum riser height shall be $7^3/_4$ inches (196 mm). The riser shall be measured vertically between leading edges of the adjacent treads. The greatest riser height within any flight of stairs shall not exceed the smallest by more than $^3/_8$ inch (9.5 mm).

R311.5.3.2 Tread depth. The minimum tread depth shall be 10 inches (254 mm). The tread depth shall be measured horizontally between the vertical planes of the foremost projection of adjacent treads and at a right angle to the tread's leading edge. The greatest tread depth within any flight of stairs shall not exceed the smallest by more than $^3/_8$ inch (9.5 mm). Winder treads shall have a minimum tread depth of 10 inches (254 mm) measured as above at a point 12 inches (305 mm) from the side where the treads are narrower. Winder treads shall have a minimum tread depth of 6 inches (152 mm) at any point. Within any flight of stairs, the largest winder tread depth at the 12 inch (305 mm) walk line shall not exceed the smallest by more than $^3/_8$ inch (9.5 mm).

Q: Is the inspection of the stair rise and run required at the framing inspection?

A: The code is not specific to this issue, although an error in this location will be a costly expense in time and money if only pointed out at the final inspection after all wall and floor finishes are in place. It takes only a moment to determine if the stair stringers will comply with code in the course of a framing inspection. In checking stairs at rough-in, the inspector will need to determine the thicknesses of proposed finish flooring materials to get accurate measurements at the top and bottom risers. If the flooring is carpeted throughout, this will not be an issue, but the use of finishes as varied as vinyl, wood flooring, quarry tile and various substrates creates significant differences. Communication between the inspector and builder is essential in this regard to avoiding costly repairs at a later date.

R311.5.3.3 Profile. The radius of curvature at the leading edge of the tread shall be no greater than $^9/_{16}$ inch (14 mm). A nosing not less than $^3/_4$ inch (19 mm) but not more than $1^1/_4$ inch (32 mm) shall be provided on stairways with solid risers. The greatest nosing projection shall not exceed the smallest nosing projection by more than $^3/_8$ inch (9.5 mm) between two stories, including the nosing at the level of floors and landings. Beveling of nosing shall not exceed $^1/_2$ inch (12.7 mm). Risers shall be vertical or sloped from the underside of the leading edge of the tread above at an angle not more than 30 degrees (0.51 rad) from the vertical. Open risers are permitted, provided that the opening between treads does not permit the passage of a 4-inch diameter (102 mm) sphere.

Exceptions:

1. A nosing is not required where the tread depth is a minimum of 11 inches (279 mm).
2. The opening between adjacent treads is not limited on stairs with a total rise of 30 inches (762 mm) or less.

Q: Is a stair nosing always required?

A: A nosing is not required where the tread depth is a minimum of 11 inches.

Q: Does the code require the risers and treads to be glued as part of the installation?

A: No. They only need to comply with the structural requirements in the code. The use of glue not only provides a better bond, but it will decrease the likelihood of squeaks in the stair.

Q: Does the requirement for an opening between treads to less than 4 inches apply to deck stairs?

A: Yes.

R311.5.4 Landings for stairways. There shall be a floor or landing at the top and bottom of each stairway.

Exception: A floor or landing is not required at the top of an interior flight of stairs, including stairs in an enclosed garage, provided a door does not swing over the stairs.

A flight of stairs shall not have a vertical rise larger than 12 feet (3658 mm) between floor levels or landings.

The width of each landing shall not be less than the width of the stairway served. Every landing shall have a minimum dimension of 36 inches (914 mm) measured in the direction of travel.

Q: We have a series of townhouses being constructed in rows. Each dwelling unit has a main entry landing that leads down a stairway that rests on a concrete landing. This landing is followed by a two-riser stair down to the sidewalk. We have reminded the builder that this concrete landing needs to be at least 36 inches in travel depth prior to meeting the two-riser stair. The builder believes that this 36-inch landing could be considered a tread, and that he should be able to construct it less than 36 inches if needed due to the location of the sidewalk. We believe that if it was less than 36 inches, and it was considered a tread, then it would not comply with code because of the nonuniformity of tread depth within $^3/_8$ inch. Are we correct in this assumption?

A: Yes. All of the wood stair treads and all of the concrete stair treads shown in Figure 3-35 would need to be uniform within $^3/_8$ inch of run.

EXTERIOR STAIR LANDINGS AT TOWNHOUSES
FIGURE 3-35

Q: A builder started the construction of a basement stairway based on the top riser meeting the vinyl floor in the kitchen on the upper level. He assumed a ¾-inch wood subfloor, ½-inch plywood underlayment and a vinyl floor as the finish material. All of the risers are at their maximum height, and they are uniform within $^3/_8$ inch. At the bottom of the stairs in the basement level the builder could barely make the required 36-inch by 36-inch landing between the bottom of the stair and the masonry foundation wall; see Figure 3-36. After the stair was framed-in, the future homeowner decided to change the kitchen floor to quarry tile. This will increase the top riser height about another 1 inch. The builder's suggestion is to add 1 inch to every riser. At the bottom landing he would like to raise it 1 inch above the concrete floor. Would this comply with code?

A: Yes, as long as the risers are uniform within $^3/_8$ inch and do not exceed the maximum rise of 7 ¾ inches. There are no provisions in the code for minimum rise of stairs or a step, and as such, the 1-inch riser at the bottom turn would comply with the code.

LANDING AT BOTTOM OF BASEMENT STAIRWAY
FIGURE 3-36

Q: Could the 1-inch riser be replaced with a little ramped area to avoid the 1-inch riser?

A: Yes.

Q: We have been told over the years that when a stairway from the main level into the basement opens into an unfinished area that a door is required at the bottom of the stairs. Is this true?

A: No.

Q: If the future homeowner wanted a door at the bottom of the stairs to the basement, what would be the minimum required landing at the bottom of the stair before the door?

A: The landing would need to be at least the width of the stair and not less than 36 inches measured in the direction of travel per Section R311.5.4. See Figure 3-37.

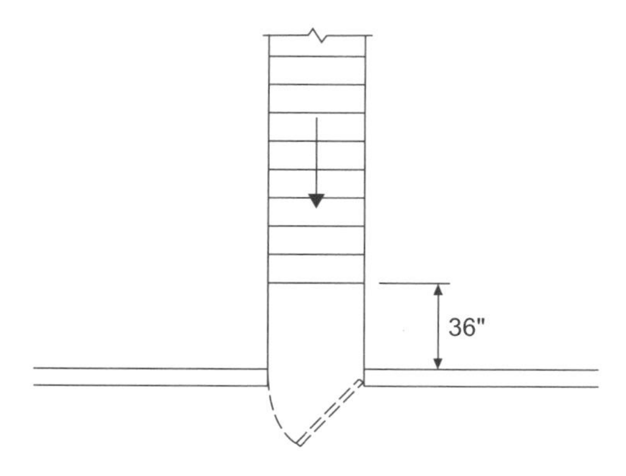

BASEMENT STAIRWAY LANDING
FIGURE 3-37

Q: Does a stairway in an attached garage require a landing at the top of the stairway?

A: Not necessarily. A landing is not required unless the door swings into the garage. See Figure 3-38.

STAIRWAY BETWEEN DWELLING AND GARAGE
FIGURE 3-38

Q: Is the walking surface of the stair tread required to be slip resistant?

A: No. Although the code addresses the maximum slope of 1 inch vertical in 48 inches horizontal, there are no provisions in the code that require a stair to be slip resistant.

R311.5.6 Handrails. Handrails shall be provided on at least one side of each continuous run of treads or flight with four or more risers.

R311.5.6.1 Height. Handrail height, measured vertically from the sloped plane adjoining the tread nosing, or finish surface of ramp slope, shall be not less than 34 inches (864 mm) and not more than 38 inches (965 mm).

R311.5.6.2 Continuity. Handrails for stairways shall be continuous for the full length of the flight, from a point directly above the top riser of the flight to a point directly above the lowest riser of the flight. Handrail ends shall be returned or shall terminate in newel posts or safety terminals. Handrails adjacent to a wall shall have a space of not less than $1^{1}/_{2}$ inch (38 mm) between the wall and the handrails.

Exceptions:

1. Handrails shall be permitted to be interrupted by a newel post at the turn.
2. The use of a volute, turnout, starting easing or starting newel shall be allowed over the lowest tread.

Q: Our department has always required the bottom newel post of a stairway to be constructed at the finish floor, or grade level. We have a situation where a deck handrail terminates over a post on the bottom tread. Is this correct?

A: In previous editions of the IRC, the handrail needed to extend to at least the bottom riser. This would normally have required the newel post to be installed at the bottom floor level. In the 2006 IRC, the language was changed to include the phrase "starting newel" over the lowest tread, which would allow the installation of the newel on the last tread. See Figure 3-39.

Q: If the starting newel is on the last tread, does it need to be placed toward the leading edge of the next tread?

A: No. The code allows it anywhere on the tread.

HANDRAIL TERMINATION
FIGURE 3-39

Q: Can the handrail extend past the starting newel?

A: The handrail ends shall be returned or shall terminate in newel posts or safety terminals per Section R311.5.6.

R311.5.7 Illumination. All stairs shall be provided with illumination in accordance with Section R303.6.

Q: The code notes that exterior illumination needs to be not less than 1 foot-candle measured at the center of treads and landings. What is 1 foot-candle, and how would it be measured?

A: There are light meters that can measure this amount of light. Although not very scientific, 1 foot-candle is approximately the amount of light given off by a full moon on a clear night.

R311.5.8 Special stairways. Spiral stairways and bulkhead enclosure stairways shall comply with all requirements of Section R311.5 except as specified below.

R311.5.8.1 Spiral stairways. Spiral stairways are permitted, provided the minimum width shall be 26 inches (660 mm) with each tread having a $7^1/_2$-inches (190 mm) minimum tread depth at 12 inches from the narrower edge. All treads shall be identical, and the rise shall be no more than $9^1/_2$ inches (241 mm). A minimum headroom of 6 feet 6 inches (1982 mm) shall be provided.

Q: Does the code require a handrail on a spiral stair to be placed on the narrower side?

A: No. The code requires a handrail on at least one side of each continuous run of treads or flight with four or more risers.

Q: We understand that Section R311.5.3.3 permits open risers, provided the openings between treads does not permit the passage of a 4-inch-diameter sphere. Does this apply to spiral stairs?

A: Yes. It applies to all stairs with a total rise of 30 inches or greater.

Q: What is the difference between a landing in a stairway and a turn in a stairway?

A: In a residential stairway, a landing is generally a 36-inch by 36-inch area where a stairway can stop, or turn direction at 90 degrees. The code has always treated stairs interrupted with a landing as two separate sets of stairs with two separate handrails. A turn is any change in direction of the stair, such as a winder stair.

Q: Does the code limit the amount of turns permitted in a stairway?

A: No.

Q: Can the continuous handrail on a stairway be interrupted by a newel post at any turn in a stairway?

A: Yes. The handrail can be interrupted by a newel post at any turn per Exception 2.

Q: Does the code limit the amount of newel posts in a stairway?

A: No. The entire stairway could be a series of turns with a newel post at each turn.

SECTION R312 GUARDS

R312.1 Guards. Porches, balconies, ramps or raised floor surfaces located more than 30 inches (762 mm) above the floor or grade below shall have guards not less than 36 inches (914 mm) in height. Open sides of stairs with a total rise of more than 30 inches (762 mm) above the floor or grade below shall have guards not less than 34 inches (864 mm) in height measured vertically from the nosing of the treads.

Porches and decks which are enclosed with insect screening shall be equipped with guards where the walking surface is located more than 30 inches (762 mm) above the floor or grade below.

R312.2 Guard opening limitations. Required guards on open sides of stairways, raised floor areas, balconies and porches shall have intermediate rails or ornamental closures which do not allow passage of a sphere 4 inches (102mm) or more in diameter.

Exceptions:

1. The triangular openings formed by the riser, tread and bottom rail of a guard at the open side of a stairway are permitted to be of such a size that a sphere 6 inches (152 mm) cannot pass through.
2. Openings for required guards on the sides of stair treads shall not allow a sphere 4 $^{3}/_{8}$ inches (107 mm) to pass through.

Q: When trying to make a determination about when a deck needs to have a guard, we have typically measured the lowest grade level within 5 feet of the deck. Is this correct?

A: In Section R312.1, a deck requires a guard if the deck surface is located more than 30 inches above the floor or grade below. That measurement is taken at the finished ground level adjoining the deck.

Q: If there is a dramatic drop in the grade just a few feet away from the deck, could our building inspectors require a guard based on that hazard?

A: There are no provisions in the code that would require a guard because of a drop in grade a few feet away from the deck.

Q: The section does not permit the passage of a 4-inch-diameter sphere. Does this apply at any location in the required guard components?

A: Yes. There are exceptions for guards at stairs in Section R312.2.

Q: It seems that insect screening could be used for a required guard in a screen porch, provided it met all the requirements for height, opening and loading. Is this correct?

A: Typically no. Section R312.2 specifically requires a complying guard at a screen porch where the walking surface is located more than 30 inches above the floor or grade below. If insect screening is proposed as any portion or component of the guard, it must be demonstrated that such components meet the loading requirements of Table R301.5 to the satisfaction of the building official. This could be very difficult, since it would be necessary to verify the structural capability of the screen as well as approve the method of attachment.

Q: We understand that the triangular section at the sides of stairs needs to be such that a sphere 6 inches in diameter cannot pass through. Why does the code allow for this less-stringent requirement?

A: The limited area created by this space and the practical difficulty in trying to meet a 4$^{3}/_{8}$-inch sphere requirement are the reasons for this provision. See Figure 3-40.

OPENING AT SIDE OF STAIRWAY
FIGURE 3-40

Q: We have a situation where a front entry door of a dwelling opens onto a sidewalk leading to the street. There is a deep window well for some basement emergency escape and rescue windows that starts at the edge of the front sidewalk; it drops approximately 48 inches. Some of our staff members believe that a guard is required only between the edge of the sidewalk where it meets the window well, and others believe the guard should extend some distance beyond the front of the window well along the grassy lawn. That opinion is based upon the possibility that somebody walking on the sidewalk might walk across the corner of the lawn and fall into the window well area. Where is a guard required?

A: For the occupants that come out of the dwelling and step onto the front sidewalk, which also serves as the front landing in this means of egress, a guard is needed directly between the window well and the entry slab. There are no sections in the code that would require a guard between the grassy area of the lawn and the window well itself. The location of the guard shown in Figure 3-41 would meet the location requirements in the code. If the builder or owner had concerns regarding this situation, he or she could certainly install a guard for the window well, keeping in mind the provisions for the escape and rescue requirements in Section R310.1.

GUARD AT WINDOW WELL
FIGURE 3-41

Q: Can a guard be installed with intermediate horizontal rails?

A: Yes. There are no provisions in the code that would prohibit the use of horizontal rails or other climbable patterns. See Figure 3-42.

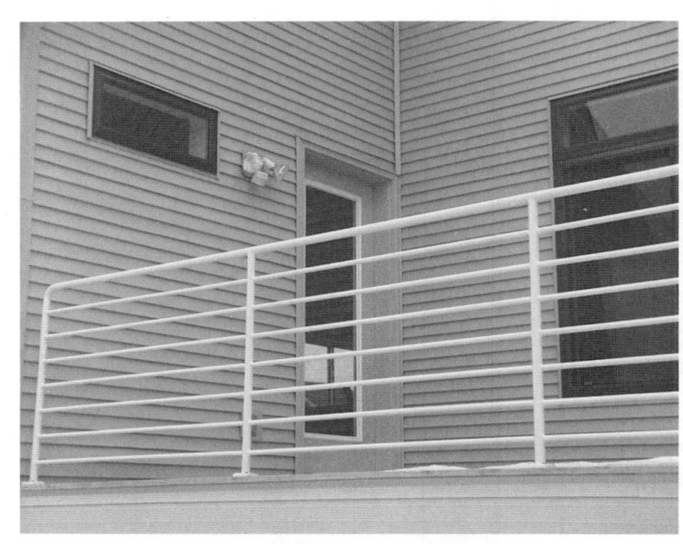

HORIZONTAL COMPONENTS IN GUARD
FIGURE 3-42

Q: Does the code contain any prescriptive requirements for how the posts that support the guard rail and components need to be installed?

A: No. There are specific loading requirements in Table R301.5 for the guardrails, handrails and guard in-fill components. The code has performance requirements, which allows for design flexibility. The top rail to the post and the post connection to the deck, historically, have been the source of many deck guard failures.

Q: Does the code allow a larger opening in guards along the side of stairways?

A: Guards on the sides of stairs need to be designed so that a $4^3/_8$-inch sphere cannot pass through at any point in the guard. See Figure 3-43.

Q: Does the code require a guard on the top of retaining walls supporting landscaping?

A: No. The guard requirements in the code apply to the porches, balconies, ramps and raised floor surfaces of the dwelling or accessory structure. There are no provisions for guards related to landscaping and retaining walls.

OPENING PROTECTION AT SIDE OF STAIRWAY
FIGURE 3-43

Q: Figure 3-44 shows a townhouse project where the front entry doors open to an elevated wood landing and then continue down a wood stairway to a concrete landing at grade. After this landing there is a two-riser stair, and another concrete landing. This landing connects to a concrete stairway down to the city sidewalk. Does the code require a guard or handrail on this concrete embankment stairway descending to the city sidewalk?

A: No. In this picture, the code addresses the front exit door, the top wood landing outside the exit door, the wood stairway and the concrete landing at the base of the wood stairway. The code does not regulate the concrete stairway on the embankment and, as such, contains no specific requirements for a guard, handrail, rise and run, or method of construction.

EMBANKMENT STAIRWAY
FIGURE 3-44

SECTION R314
FOAM PLASTIC

R314.1 General. The provisions of this section shall govern the materials, design, application, construction and installation of foam plastic materials.

R314.2 Labeling and identification. Packages and containers of foam plastic insulation and foam plastic insulation components delivered to the job site shall bear the label of an approved agency showing the manufacturer's name, the product listing, product identification and information sufficient to determine that the end use will comply with the requirements.

R314.3 Surface burning characteristics. Unless otherwise allowed in Section R314.5 or R314.6, all foam plastic or foam plastic cores used as a component in manufactured assemblies used in building construction shall have a flame spread index of not more than 75 and shall have a smoke-developed index of not more than 450 when tested in the maximum thickness intended for use in accordance with ASTM E 84. Loose-fill-type foam plastic insulation shall be tested as board stock for the flame spread index and smoke-developed index.

> **Exception:** Foam plastic insulation more than 4 inches thick shall have a maximum flame spread index of 75 and a smoke-developed index of 450 where tested at a minimum thickness of 4 inches, provided the end use is approved in accordance with Section R314.6 using the thickness and density intended for use.

R314.4 Thermal barrier. Unless otherwise allowed in Section R314.5 or Section R314.6, foam plastic shall be separated from the interior of a building by an approved thermal barrier of minimum 0.5 inch (12.7 mm) gypsum wallboard or an approved finish material equivalent to a thermal barrier material that will limit the average temperature rise of the unexposed surface to no more than 250°F (139°C) after 15 minutes of fire exposure complying with the ASTM E 119 standard time temperature curve. The thermal barrier shall be installed in such a manner that it will remain in place for 15 minutes based on NFPA 286 with the acceptance criteria of Section R315.4, FM 4880, UL 1040 or UL 1715.

Q: The code requires all foam plastic used in building construction to have a flame spread index of not more than 75. What does the 75 refer to?

A: The flame-spread index is based on ASTM E 84. This standard is based on the "Steiner Tunnel Test," where the material is tested in a 25-foot-long tunnel that measures the ability of a material to carry a flame. The base standard of 100 is the flame spread rating of red oak wood. If a foam plastic insulation had a flame spread index of 75, then it would literally burn 75 percent as fast as red oak.

Q: How would the user of the foam plastic know what the flame spread and smoke-developed rating is?

A: Section R314.2 requires that information to be on the label of the product.

Q: Section R314.4 requires foam plastic insulation to be separated from the interior of the building by a minimum ½-inch gypsum board or approved finish material equivalent to a 15-minute thermal barrier. Who determines what is equivalent? See Figure 3-45.

A: The building official makes that determination based on supporting data submitted by the proponent of the foam product.

SPRAY-APPLIED FOAM INSULATION
FIGURE 3-45

Q: What kind of data?

A: A proponent (owner/designer/builder) would need to submit sufficient data to the building official for consideration when an alternative material is proposed that is not specifically referenced in the IRC. The material would need to be tested per ASTM E 119, and be shown to limit the average temperature rise for 15 minutes as specified in the standard.

Q: What is ASTM E 119?

A: ASTM E 119 is the *Standard Test Methods for Fire Tests of Building Construction and Materials* established by ASTM International (previously known as the American Society for Testing and Materials).

Q: What does this standard actually measure?

A: This standard is used to measure and describe the response of materials, products or assemblies to heat and flame under controlled conditions. This test prescribes a standard fire exposure for comparing the test results of building materials, which will be used as a factor in determining the fire performance of a material or assembly during fire conditions.

Q: Are there other sources available to determine thermal barriers?

A: Yes. In Section 720 of the IBC, there is prescriptive information that can be used to evaluate the fire performance of building materials. For example, in Table 720.6.2(1), under "Calculated Fire Resistance," it rates $^{19}/_{32}$-inch structural wood panels bonded with exterior glue (Exposure 1 rating) to have a 15-minute fire rating. The APA- Engineered Wood Association, has a Technical Topic number 060A, *APA Performance Rated Wood Structural Panels as Thermal Barriers,* that lists $^{23}/_{32}$-inch Douglas fir plywood as being 15 minute rated when tested to ASTM E 119. For example, if a proponent wanted to install foam plastic insulation in the exterior stud walls of his or her dwelling, it would allow the use of one of these tested wood structural panels as the required protection versus the installation of ½-inch gypsum board.

Q: Can a flame-retardant poly vapor barrier serve as the thermal barrier to protect a foam insulation product?

A: No. A flame-retardant poly is considered to be flame retardant because the flame spread rating (per ASTM E 84) is less than 25, which basically allows the poly itself to remain unprotected by a thermal barrier. In order for a product to serve as a 15-minute thermal barrier, it needs to be tested per ASTM E 119. The poly most likely is not tested to ASTM E 119.

Q: Figure 3-46 shows a rigid foam plastic insulation in a rim joist area. The builder does not intend to cover this foam with any thermal barrier because of the flame spread and smoke-developed rating of the foam. Can we approve this installation?

A: Section R314.6 contains criteria for testing of foam plastic insulation products that may allow the foam to remain unprotected without a thermal barrier. The specific approval is based on the actual end-use configuration of the finished assembly in the maximum thickness intended for use. The test results or possibly an ICC ES report would most likely be the basis for the building official to approve the installation.

RIGID FOAM PLASTIC INSULATION
FIGURE 3-46

Q: Can a construction adhesive be used to fasten the ½-inch gypsum board over the foam plastic insulation?

A: The construction adhesive would need to be tested for the specific application as noted in Section R314.4.

R314.5.3 Attics. The thermal barrier specified in Section 314.4 is not required where attic access is required by Section R807.1 and where the space is entered only for service of utilities and when the foam plastic insulation is protected against ignition using one of the following ignition barrier materials:

1. 1.5-inch-thick (38 mm) mineral fiber insulation;
2. 0.25-inch-thick (6.4 mm) wood structural panels;
3. 0.375-inch (9.5 mm) particleboard;
4. 0.25-inch (6.4 mm) hardboard;
5. 0.375-inch (9.5 mm) gypsum board; or
6. Corrosion-resistant steel having a base metal thickness of 0.016 inch (0.406 mm).

The above ignition barrier is not required where the foam plastic insulation has been tested in accordance with Section R314.6.

Q: When the code refers to "service of utilities" in an attic or crawl space, what type of utilities does it address?

A: These utilities include heating and air-conditioning equipment, wiring, ducts and similar equipment.

Q: A builder has installed one layer of ½-inch gypsum board on the bottom side of the roof trusses. From the inside of the attic he would like to install spray-applied foam insulation 8 inches thick onto the gypsum board to serve as the attic insulation, and then cover it with 1½ inches of cellulose. If the attic is only going to be accessed for service of utilities, would this situation comply with the code?

A: No. There are no provisions for the use of cellulose insulation as an ignition barrier in Section R314.5.3. The builder would need to cover the foam with one of the specific materials listed. Also, the spray-applied foam cannot be installed greater than 4 inches in thickness unless it has been shown to meet specific requirements and testing of Section R314.6. See Figure 3-47.

Q: What kind of testing does this involve?

A: Section R314.6 addresses testing and acceptance criteria for the installation of foam in thickness greater than 4 inches. The approved tests are NFPA 286, FM 4880, UL 1040 or UL 1715, or fire tests related to actual end-use configurations.

Q: Could this testing also address alternative materials for ignition barriers, or possibly the use of a product without an ignition barrier?

A: Yes, it is possible. The results of the testing would establish how it can be installed.

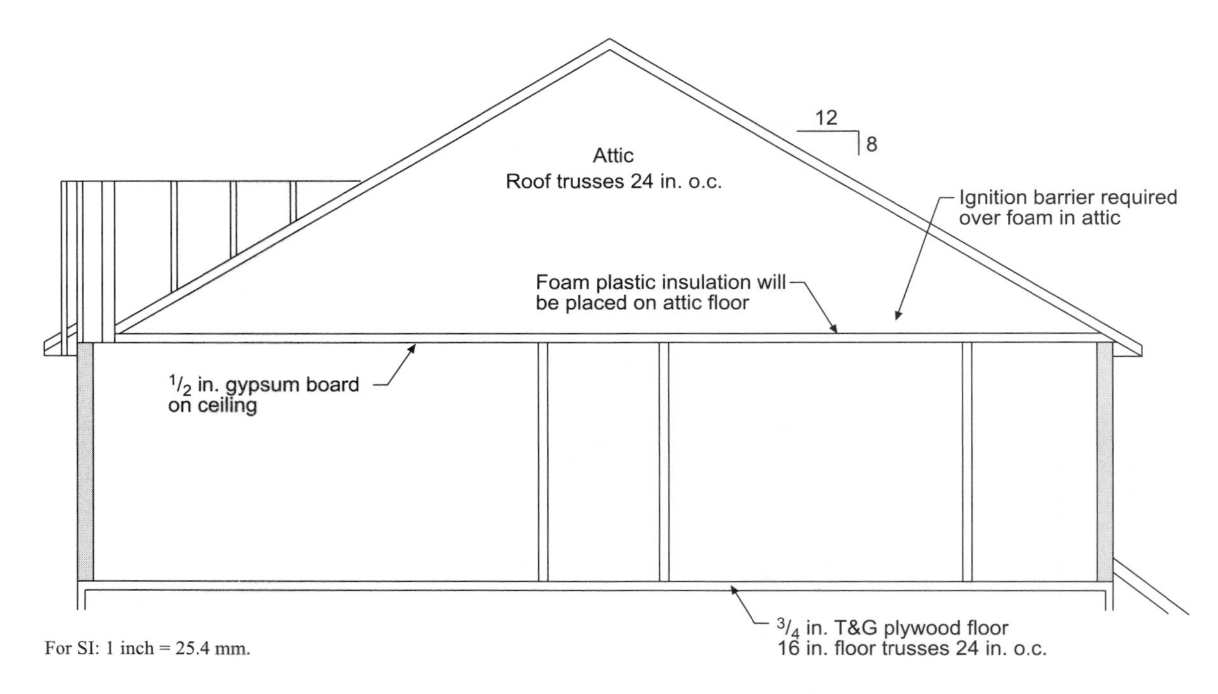

FOAM PLASTIC INSULATION IN ATTIC
FIGURE 3-47

Q: Section R314.5.3, which addresses attics, states that the thermal barrier is not required where attic access is required and where the space is entered only for service of utilities, provided the foam plastic insulation is protected by an ignition barrier. Does this mean that if there is no attic access, then there are no requirements to provide a thermal barrier?

A: No. Section R314.5.3 allows a reduction of the 15-minute thermal barrier to prevent accidental infringement of flames on foam plastic in areas not normally used by occupants when accessed only for service of utilities. There are six ignition barrier materials noted in the section that can be used to meet this requirement.

Q: What if the attic area was used for storage of household goods?

A: If the attic area was used for stored materials, it would no longer be considered an area where the space is entered only for service of utilities. The code would then require ½-inch gypsum board or an equivalent thermal barrier to separate the foam plastic insulation from the rest of the attic space.

R807.1 Attic access. Buildings with combustible ceiling or roof construction shall have an attic access opening to attic areas that exceed 30 square feet (2.8 m^2) and have a vertical height of 30 inches (762 mm) or more.

Q: Is an attic access required?

A: An attic access is required by Section R807.1 if certain conditions exist.

R314.5.4 Crawl spaces. The thermal barrier specified in Section R314.4 is not required where crawlspace access is required by Section R408.3 and where entry is made only for service of utilities and the foam plastic insulation is protected against ignition using one of the following ignition barrier materials:

1. 1.5-inch-thick (38 mm) mineral fiber insulation;
2. 0.25-inch-thick (6.4 mm) wood structural panels;
3. 0.375-inch (9.5 mm) particleboard;
4. 0.25-inch (6.4 mm) hardboard;
5. 0.375-inch (9.5 mm) gypsum board; or
6. Corrosion-resistant steel having a base metal thickness of 0.016 inch (0.41 mm).

The above ignition barrier is not required where the foam plastic insulation has been tested in accordance with Section R314.6.

Q: A plan has been submitted that indicates a rigid foam plastic insulation on the inside face of the exterior foundation wall of a crawl space. What protection is required per the code?

A: Section R314.5.4 indicates materials that can be used to protect the foam where entry is made only for service of utilities. If the crawl space is intended to be used for some other purpose, such as storage of materials, ½-inch gypsum board or an equivalent 15-minute barrier needs to be installed. Also, the foam insulation in the figure would not provide a continuous insulation barrier. See Figure 3-48.

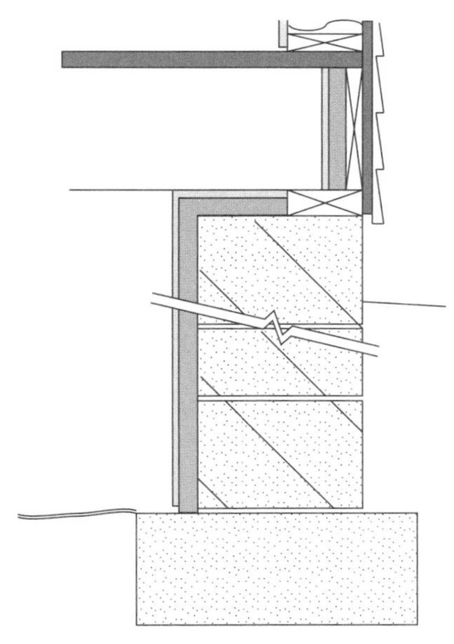

FOAM PLASTIC INSULATION IN CRAWL SPACE
FIGURE 3-48

R314.5.5 Foam-filled exterior doors. Foam-filled exterior doors are exempt from the requirements of Sections R314.3 and R314.4.

R314.5.6 Foam-filled garage doors. Foam-filled garage doors in attached or detached garages are exempt from the requirements of Sections R314.3 and R314.4.

Q: Why are foam-filled exterior doors and foam-filled garage doors exempt from the requirements related to flame spread, smoke-developed rating and thermal barrier protection?

A: The exemption is based on the lack of fire hazard related to the ignition of these doors on the outside of the structure.

R314.5.12 Sheathing. Foam plastic insulation used as sheathing shall comply with Section R314.3 and Section R314.4. Where the foam plastic sheathing is exposed to the attic space at a gable or kneewall, the provisions of Section R314.5.3 shall apply.

Q: Section R314.5.12 appears to allow foam sheathing on the exterior of the house to be exempt from the requirements related to flame spread, smoke-developed rating and thermal barrier as long as it is not exposed in the attic or kneewall. Is this correct?

A: No. The foam sheathing needs to comply with the requirements in Section R314.3 for surface-burning characteristics and Section R314.4 for thermal barrier, which addresses foam separated from the interior of the building. See Figure 3-49.

EXTERIOR FOAM PLASTIC INSULATION
FIGURE 3-49

Q: Can foam sheathing installed on the exterior of a house be used as a weather-resistive barrier material?

A: Although many rigid foam plastic insulation products are tested for flame spread and smoke-developed ratings, many are not tested for application as the weather-resistive. The manufacturer of the foam plastic should be able to provide data to indicate how the product can be installed. The weather-resistive barrier requirements are contained in Section R703.

Q: We have a question related to spray-applied foam plastic insulation, although it is not an energy code question. We have a situation where a 4-inch thick layer of spray-applied foam plastic has been installed in a 2 x 6 stud cavity. It was installed to the minimum requirement based upon the necessary insulation value required by the energy code for our geographical area. The stud wall has a single plate at the bottom and a double plate at the top. The builder intends to install ½-inch gypsum board on the inside face of the wall as the finish material. Although the foam in the stud cavity is limited by 2 x 6 studs on each side and 2 x 6 plates top and bottom, will the application of the gypsum board meet the requirements for the thermal barrier even though there is about a 1½-inch gap between the foam and the gypsum board?

A: Although the code does not address the situation of the 1 ½-inch gap in the stud cavity between the foam and the interior gypsum board, it would appear that the ½-inch gypsum board would meet the intent of the separation from the interior in this situation. See Figure 3-50.

FOAM PLASTIC INSULATION IN STUD WALL
FIGURE 3-50

Q: A builder has installed a spray-applied foam on the outside of the concrete masonry foundation wall that extends from the top of the footing to a point approximately 8 inches above grade; see Figure 3-51. We have concerns about this installation and we are not sure where to start. Any suggestions?

A: Start with the manufacturer's installation instructions to give you the information necessary to determine if this spray-applied foam can be installed on the exterior, installed above grade without protection from physical damage and installed with some other product sprayed onto it (compatibility of products).

Q: Could this installation of spray-applied foam on the concrete serve as a dampproofing or waterproofing installation?

A: You may need to review appropriate test data and the manufacturer's installation instructions for both of the products in Figure 3-51 to determine how these products can be installed. It is possible that the particular foam product may not be manufactured for ground contact, or exposure to the weather extremes. Some foam products need to be protected for exposure from ultraviolet rays of the sun that can break down the material. If available, an ICC ES report would most likely address these installation issues.

SPRAY-APPLIED FOAM PLASTIC INSULATION
FIGURE 3-51

R314.5.11 Sill plates and headers. Foam plastic shall be permitted to be spray applied to a sill plate and header without the thermal barrier specified in Section R314.4 subject to all of the following:

1. The maximum thickness of the foam plastic shall be $3^1/_4$ inches (83 mm).
2. The density of the foam plastic shall be in the range of 1.5 to 2.0 pounds per cubic foot (24 to 32 kg/m^3).
3. The foam plastic shall have a flame spread index of 25 or less and an accompanying smoke developed index of 450 or less when tested in accordance with ASTM E 84.

Q: Section R314.5.11, "Sill plates and headers," allows the use of unprotected spray-applied foam plastic meeting the criteria of the section to remain unprotected. Can this provision be applied to 2 x 4 walls?

A: No. This provision in the code is based on a specific modified room corner test (with a wood floor joist system) conducted with a spray-applied foam plastic product meeting the three requirements in the section. This provision only applies to spray-applied foam in a rim joist area, also known as sill plate and header. See Figure 3-52.

Q: Could this provision apply to rigid foam plastic insulation?

A: No. It only applies to spray-applied foam plastic insulation.

SPRAY-APPLIED FOAM IN RIM JOIST/SILL PLATE AND HEADER AREA
FIGURE 3-52

Q: Why does the provision contain a maximum thickness of 3¼ inches?

A: The testing data was based on that thickness.

Q: The density of the foam plastic insulation according to the section appears to address the use of closed-cell foam plastic insulation. Would this prohibit the use of open-cell foam plastic insulation?

A: In order to use the reduction of the thermal barrier noted in Section R314.5.11, the foam will need to meet the density listed in Item 2, that being between 0.5 and 2.0 pounds per cubic foot (pcf).

Q: Although not specifically a code question, our inspectors have heard conflicting information regarding the issue of density of foam. In order to approve the products, we are trying to verify compliance with the builders and the contractors that are spraying the foam in the field. We have heard that some of the foam-blowing agents have been phased out. How does this affect the foam, and is this a concern to us as code officials?

A: In response to the United States Clear Air Act (www.epa.gov), certain substances called hydrochlorofluorocarbons (HCFCs), such as HCFC-141b and others used for foam blowing, are being phased out. This change in spraying agents (aerosols) may affect the density of the foam plastic insulation products. The manufacturers of the foam should be able to provide updated specifications as requested.

Q: Is there any criteria for a spray-applied foam plastic not meeting the density of 1.5 to 2.0 pcf to be applied in this sill plate and header (rim joist) location unprotected by a thermal barrier?

A: Yes. It would need to meet the test criteria of Section R314.6, based on the end-use conditions.

R314.6 Specific approval. Foam plastic not meeting the requirements of Sections R314.3 through R314.5 shall be specifically approved on the basis of one of the following approved tests: NFPA 286 with the acceptance criteria of Section R315.4, FM4880, UL 1040 or UL 1715, or fire tests related to actual end-use configurations. The specific approval shall be based on the actual end use configuration and shall be performed on the finished foam plastic assembly in the maximum thickness intended for use. Assemblies tested shall include seams, joints and other typical details used in the installation of the assembly and shall be tested in the manner intended for use.

Q: Are there any provisions in the code that would allow the manufacturer of a rigid foam plastic insulation to remain unprotected by a thermal barrier in other locations within the dwelling?

A: Yes. Section R314.6 contains tests based on end-use conditions.

Q: Our department is aware that some rigid foam plastic insulation products have completed special fire testing where the results indicate that the product can be left unprotected anywhere in a dwelling. Does the foil face on these products have anything to do with passing these fire tests?

A: No. The foil does not provide any significant thermal protection of the foam. The favorable test results are based on the performance of the foam without regard to the foil.

Q: Why is this foil applied to the foam plastic?

A: The foil usually serves as a means to control the thickness and flatness of the foam in the production of the product.

Q: Would the foil serve as a vapor retarder?

A: It could, based on the thickness of the foil and the permeability rating.

Q: Do foam plastic insulation products need to meet the "critical radiant flux" requirements for attic insulation noted in Section R316.4?

A: No. The critical radiant flux requirements referenced in Sections R316.4 and R316.5, based on ASTM E 970, apply to attic insulation products such as cellulose and mineral fiber. They do not apply to foam plastic insulation.

Q: Can foam plastic insulation be installed in contact with plastic-based pipe/vents?

A: The code does not address the issue. The user should refer to the manufacturer's installation instructions for both the foam and the pipe. See Figure 3-53.

FOAM PLASTIC INSULATION AT VERTICAL PIPE PENETRATION
FIGURE 3-53

SECTION R315
FLAME SPREAD AND SMOKE DENSITY

R315.1 Wall and ceiling. Wall and ceiling finishes shall have a flame-spread classification of not greater than 200.

> **Exception:** Flame-spread requirements for finishes shall not apply to trim defined as picture molds, chair rails, baseboards and handrails; to doors and windows or their frames; or to materials that are less than $^1/_{28}$ inch (0.91 mm) in thickness cemented to the surface of walls or ceilings if these materials have a flame-spread characteristic no greater than paper of this thickness cemented to a noncombustible backing.

R315.2 Smoke-developed index. Wall and ceiling finishes shall have a smoke-developed index of not greater than 450.

R315.3 Testing. Tests shall be made in accordance with ASTM E 84.

Q: How is the flame spread and smoke-developed rating of an interior trim material determined and verified?

A: Refer to the manufacturer's specifications. In Section R315.1, there are specific exceptions for some particular interior finish materials.

SECTION R317
DWELLING UNIT SEPARATION

R317.1 Two-family dwellings. Dwelling units in two-family dwellings shall be separated from each other by wall and/or floor assemblies having not less than a 1-hour fire-resistance rating when tested in accordance with ASTM E 119. Fire-resistance-rated floor-ceiling and wall assemblies shall extend to and be tight against the exterior wall, and wall assemblies shall extend to the underside of the roof sheathing.

Exceptions:

1. A fire-resistance rating of $^1/_2$ hour shall be permitted in buildings equipped throughout with an automatic sprinkler system installed in accordance with NFPA 13.
2. Wall assemblies need not extend through attic spaces when the ceiling is protected by not less than $^5/_8$-inch (15.9 mm) Type X gypsum board and an attic draft stop constructed as specified in Section R502.12.1 is provided above and along the wall assembly separating the dwellings. The structural framing supporting the ceiling shall also be protected by not less than $^1/_2$ -inch (12.7 mm) gypsum board or equivalent.

R317.1.1 Supporting construction. When floor assemblies are required to be fire-resistance-rated by Section R317.1, the supporting construction of such assemblies shall have an equal or greater fire-resistive rating.

Q: Our department believes that a two-family dwelling could be described as two dwelling units in the same structure located on the same lot. Is this correct?

A: Yes. A two-family dwelling has no lot line separating the units.

Q: We understand that the separation needs to be at least 1-hour fire-resistance rated. What is ASTM E 119?

A: ASTM E 119 is the *Standard Test Method For Fire Tests Of Building Construction and Materials.* The ASTM E 119 test measures the performance of a designed wall, floor or roof assembly, related to excessive temperature rise, passage of flame or structural collapse.

Q: It appears the code would require only one fire-resistance-rated wall, with a 1-hour rating, to be installed between the two dwelling units basically from front to back. Would this 1-hour fire-resistance-rated wall need to extend to the front of the structure that also separates the two attached garages in front?

A: Yes. There are no provisions that would allow a reduction in the fire-resistance rating specifically between the two garages.

Q: Is there a parapet required for this two-family dwelling separation?

A: No, provided the 1-hour fire-resistance-rated wall extends up through the attic to the roof sheathing and out into the soffits at the overhang. The parapet provisions only apply to townhouses.

Q: Section R317.1 would also allow the 1-hour fire-resistance-rated assembly to extend up to the ceiling, requiring only draftstopping in the attic, as long as the ceiling was protected with not less than one layer of $^5/_8$-inch Type X gypsum board. Would this ceiling need to meet ASTM E 119?

A: No. Only the 1-hour fire-resistance-rated wall between the units would need to meet ASTM E 119.

Q: If it does not need to meet ASTM E 119, then do the penetrations in the ceiling, such as those for electrical, mechanical and plumbing, need to meet the requirements in Section R317.3 for rated penetrations?

A: No, only penetrations of rated fire-resistant wall and floor/ceiling assemblies must comply with Section R317.3.

Q: With the average age of the population of our area increasing, our department has seen an increase in situations where a two-family dwelling is constructed where the parents live in the dwelling unit on one side and the family of a son or daughter lives on the other side, only separated by the 1-hour rated fire-resistant wall. The son or daughter would like to provide the independence for their parents next door, but also have the availability of entry between the units if needed on very short notice. Can a fire-rated door be installed in the 1-hour fire-resistance-rated wall between the units and still comply with the code?

A: Although this situation of side-by-side dwelling units in one dwelling is not addressed in the IRC, a building official could consider the provisions for openings between dwelling units in the IBC that require a 45-minute fire-resistance-rated door assembly in a 1-hour fire-resistance-rated wall when evaluating this condition. This answer would not be valid if a lot line was platted between the two dwelling units because then they would need to meet the requirements of Section R302.1 based on location on lot as separate dwellings.

Q: When a two-family dwelling is constructed as an upper/lower layout with a 1-hour fire-resistance-rated floor assembly, are openings permitted in this 1-hour fire-resistance-rated floor assembly that separates the dwelling units for ductwork? See Figure 3-54.

A: Openings for ductwork, such as warm air supply vents and cold air return vents, can only be installed if they are included in the listing of the assembly. Some listed assemblies, such as those noted in the UL *Fire Resistance Directory,* are very specific as to what openings or penetrations are permitted in the horizontal assembly based on the testing.

Q: If a horizontal 1-hour fire-resistance-rated assembly is installed, are the supporting walls required to be rated?

A: Yes. Section R317.1.1 requires the supporting walls to have an equal or greater fire-resistive rating.

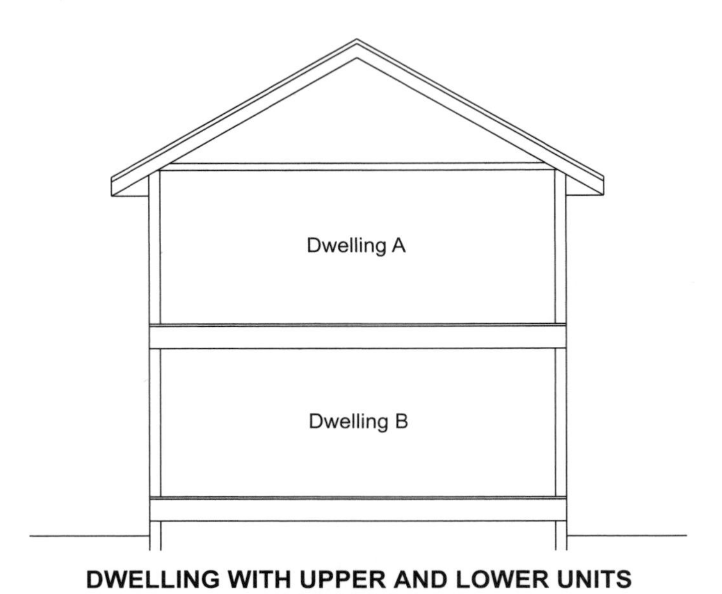

DWELLING WITH UPPER AND LOWER UNITS
FIGURE 3-54

Q: Are windows permitted in these rated walls that support the horizontal floor/ceiling assembly?

A: Yes, as long as the openings are not prohibited by some other section of the code, such as in Table R302.1, related to location on lot. The purpose of the fire-resistant rating is to protect the structural integrity of the wall supporting the rated ceiling, not to prevent fire from passing through the openings.

Q: If this side-by-side dwelling unit in Figure 3-55 was constructed with a lot line between the dwellings, how do the provisions in Section R317.1 for two-family dwellings apply?

A: Section R317.1 does not apply to two dwellings separated by a lot line. These two separate dwelling units would require protection based on the location on lot. Refer to the Q&A in Section R302.1 for exterior walls.

Q: Are residential sprinklers required for a townhouse?

A: No. There are no requirements for residential sprinklers in the IRC.

R317.2 Townhouses. Each townhouse shall be considered a separate building and shall be separated by fire-resistance-rated wall assemblies meeting the requirements of Section R302 for exterior walls.

R302.1 Exterior walls. Construction, projections, openings and penetrations of exterior walls of dwellings and accessory buildings shall comply with Table R302.1. These provisions shall not apply to walls, projections, openings or penetrations in walls that are perpendicular to the line used to determine the fire separation distance. Projections beyond the exterior wall shall not extend more than 12 inches (305 mm) into the areas where openings are prohibited.

Q: Where is the fire separation requirement for townhouses located in the code?

A: The provisions for the fire separation of townhouses are contained in Section R317.2 for townhouses and Section R302.1 for exterior walls. The IRC provisions for townhouses are based upon each dwelling unit being treated as a separate building, and the

TABLE R302.1
EXTERIOR WALLS

EXTERIOR WALL ELEMENT		MINIMUM FIRE-RESISTANCE RATING	MINIMUM FIRE SEPARATION DISTANCE
Walls	(Fire-resistance rated)	1 hour with exposure from both sides	0 feet
	(Not fire-resistance rated)	0 hours	5 feet
Projections	(Fire-resistance rated)	1 hour on the underside	2 feet
	(Not fire-resistance rated)	0 hours	5 feet
Openings	Not allowed	N/A	< 3 feet
	25% Maximum of Wall Area	0 hours	3 feet
	Unlimited	0 hours	5 feet
Penetrations	All	Comply with Section R317.3	< 5 feet
		None required	5 feet

N/A = Not Applicable.

separating walls are considered exterior walls on a lot line.

Q: If a townhouse is constructed using the provisions in Table R302.1 for exterior walls, what is the minimum requirement for the fire-resistance rating of the common exterior walls between the dwellings?

A: A townhouse is "considered a separate building," per Section R317.2. The requirement in Table R302.1 for exterior walls addresses the distance of the exterior wall of each dwelling unit in relation to the fire separation distance. If the exterior wall has a fire separation distance of less than 5 feet as noted in the table, then the exterior wall needs to be constructed with a minimum fire-resistance rating of 1-hour. This would apply to each of the adjacent dwelling units. Because each dwelling unit has a fire separation distance less than 5 feet per Table R302.1, each dwelling unit's exterior wall that is parallel to the lot line (or imaginary lot line) needs to be 1-hour minimum fire-resistance rated with exposure from both sides.

Q: What does "exposure from both sides" refer to as far as the fire-resistance rating?

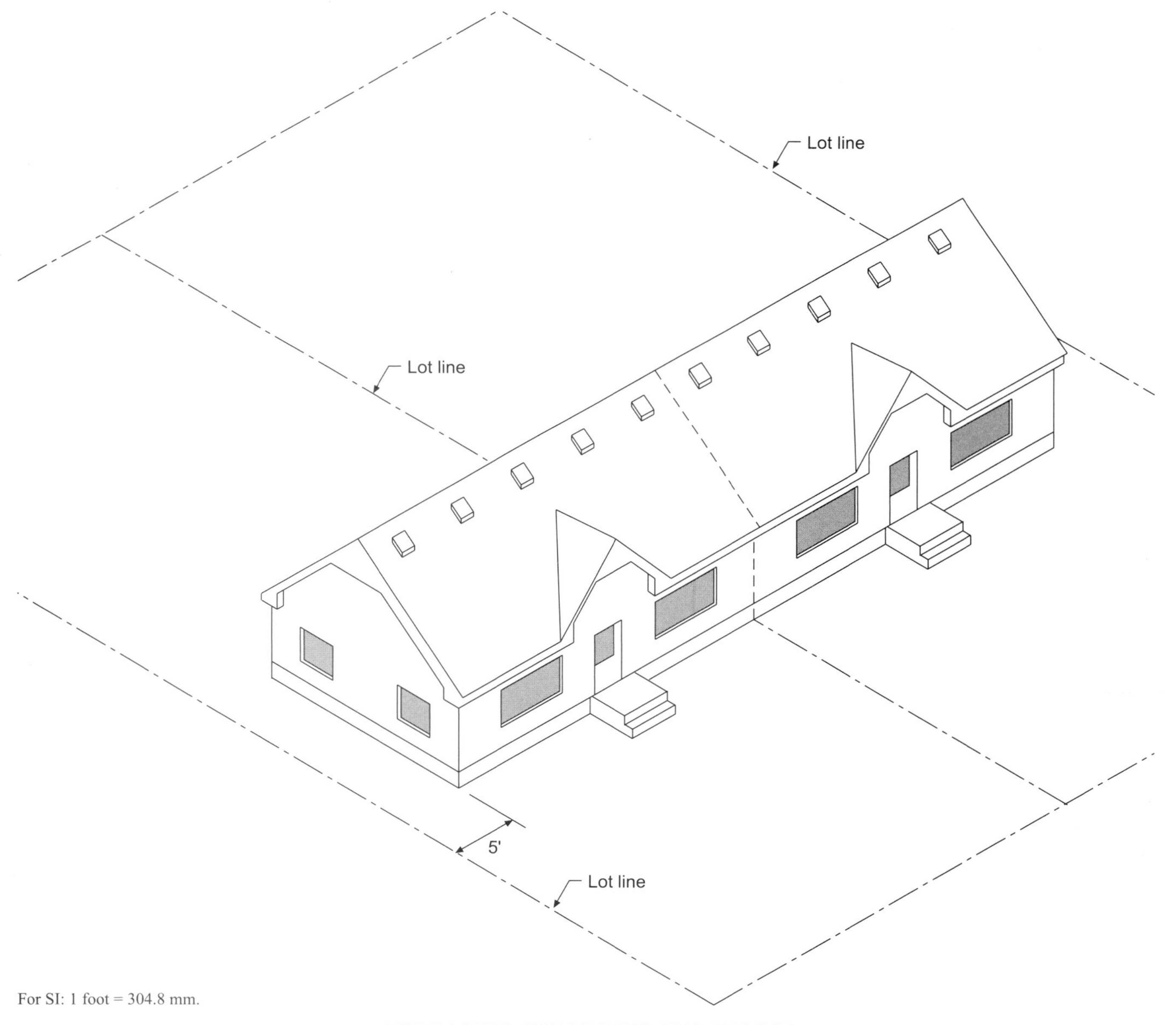

For SI: 1 foot = 304.8 mm.

DWELLINGS SEPARATED BY LOT LINE
FIGURE 3-55

A: The fire-resistance rating is based on fire testing from both sides.

Q: Is the fire separation distance a measurement between the exterior wall and a lot line?

A: This could be the distance to a lot line, but there are other options. In Section R202, "Definitions," it defines "Fire separation distance" as follows:

FIRE SEPARATION DISTANCE. The distance measured from the building face to one of the following:

1. To the closest interior lot line; or
2. To the centerline of a street, an alley or public way; or
3. To an imaginary line between two buildings on the lot.

The distance shall be measured at a right angle from the face of the wall.

Q: Could a single 2-hour fire-resistance-rated wall be used for the fire separation between two dwelling units in a townhouse instead of the two 1-hour fire-resistance-rated walls?

A: Yes. A 2-hour fire-resistance-rated wall could be installed per the exception to Section R317.2, "Townhouses," that states, "A common 2-hour fire-resistance-rated wall is permitted for townhouses if such walls do not contain plumbing or mechanical equipment, ducts or vents in the cavity of the common wall."

Q: In the course of a plan review of a townhouse, what information should be provided to the building official to determine if the designated fire-resistance-rated walls comply with ASTM E 119?

A: Documentation of a tested or listed assembly should be provided at the time of plan review. Rated assemblies contained in recognized publications, such as the *Fire Resistance Directory* published by Underwriters Laboratory, or the *Fire Resistance Design Manual* published by the U.S. Gypsum Association, or other approved documents by the building official could be considered.

Q: A designer has submitted a design for a listed/tested 2-hour fire-resistance-rated wall assembly to be used for the separation between the dwelling units of a townhouse project. He would like to add insulation to the wall cavity of this assembly to increase the sound rating. Can this be done without affecting the assembly?

A: The particular listed assembly should be verified with the listing agency. For example, for assemblies listed in the U.S. Gypsum Association *Fire Resistance Design Manual*, it states, "When not specified as a component of a fire tested wall or partition system, mineral fiber, glass fiber, or cellulose fiber insulation of a thickness not exceeding that of the stud depth shall be permitted to be added within the stud assembly." In Section 721.6.2.5 of the IBC, "Additional protection," it contains a related table with time increments that can be added to the fire resistance where glass fiber, rockwool, slab mineral wool or cellulose insulation is incorporated in the assembly.

R317.2.1 Continuity. The fire-resistance-rated wall or assembly separating townhouses shall be continuous from the foundation to the underside of the roof sheathing, deck or slab. The fire-resistance rating shall extend the full length of the wall or assembly, including wall extensions through and separating attached enclosed accessory structures.

Q: Figure 3-56 shows a four-unit townhouse with attached garages in front of each unit. There are lot lines between each unit extending front to back. The front of the dwelling is accessed by a walkway that travels past the side of the garages and up to the front of the dwelling units. When referring to "continuity," Section R317.2.1 also states, "and separating attached enclosed accessory structures." Would the attached garages in our drawing be considered an "accessory structure" under this provision?

A: Yes. The two 1-hour fire-resistance-rated walls between the garages of Unit B and Unit C would need to extend between the garages and extend out to the front of the garages. The garages of Unit A and Unit D do not require any additional protection because of their location on the lot.

Q: Figure 3-57 shows a townhouse project where the decks are literally continuous along the front of the structure, only separated by a nonfire-resistance-rated wall that is there for visual separation. We understand the 2-hour fire separation needs to extend to the exterior wall (or overhang) of the structure, but does it need to extend between these attached exterior decks?

A: No. A deck is not "enclosed," such as an attached garage or accessory structure, and thus requires no fire-resistant-rated wall separation between the decks of each dwelling unit in the townhouse project in your photograph.

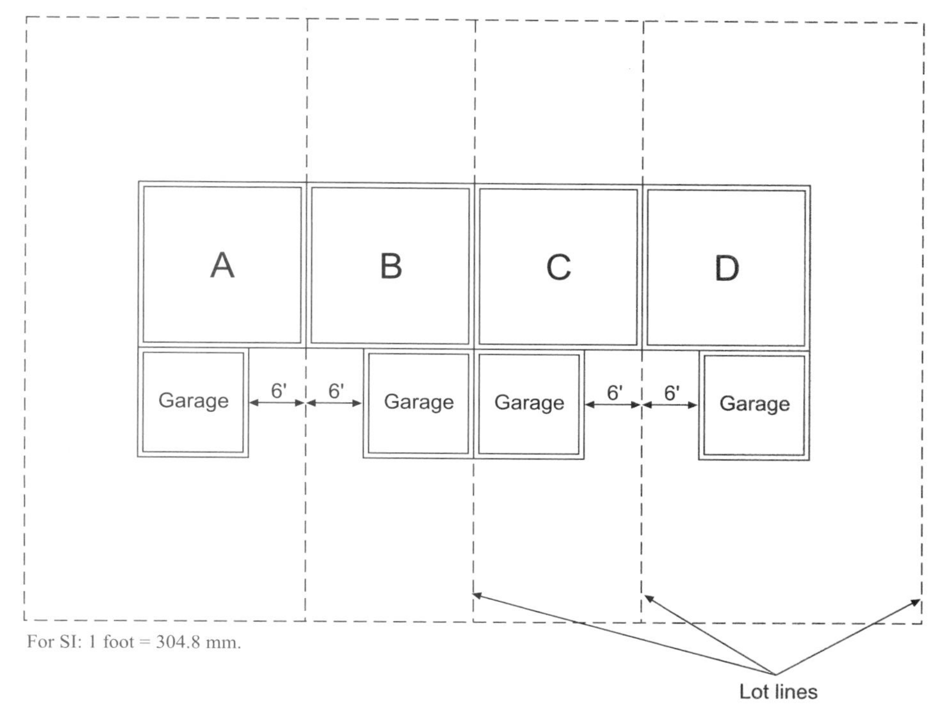

For SI: 1 foot = 304.8 mm.

ATTACHED GARAGES IN FRONT OF DWELLING UNITS IN TOWNHOUSE
FIGURE 3-56

SIDE-BY-SIDE UNITS IN A TOWNHOUSE
FIGURE 3-57

Q: We would like your opinion on the townhouse in Figure 3-58. It is understood that the project meets the definition of a "Townhouse" because each unit has two sides open. Units A, D, E and F all have a front and side open. Units B and C have a front and back open. We understand there are two hours of fire separation between Units A and B. We have two questions related to this layout. First, does the exterior wall between Unit E and the open lot serving Unit B (indicated by lot line "X" in the drawing) need to be 2-hour fire-resistance rated or 1-hour fire-resistance rated?

A: It needs to be 1-hour fire-resistance rated due to the location on lot (less than 5 feet from the lot line).

Q: Second, can this same wall between Unit E and the open lot serving Unit B have any windows or doors?

A: No. Openings are not permitted due to location on lot (less than 3 feet from the lot line).

Q: We have a question regarding the separation of a deck attached to a dwelling unit in a townhouse and the adjacent dwelling unit. In Figure 3-59, this townhouse is made up of three attached dwelling units (Units A, B and C). There is a lot line that extends between Unit A and Unit B that extends out to the public way. The common wall between Unit A and Unit B is designed as two 1-hour fire-resistance-rated assemblies. Unit B projects beyond the front and rear of Unit A and Unit C. Does the exterior wall of Unit B that extends beyond the rear of Unit A need to be a fire-resistance-rated wall?

A: Yes. This exterior wall of Unit B is required to be at least 1-hour fire-resistance rated based on the fire separation distance from the lot line between Unit A and Unit B that extends out to the public way. The deck shown does not affect the answer. It is the same with the deck located per the drawing and without the deck.

Q: Figure 3-60 has been submitted for a four-unit townhouse project. The four units are in a row. The front entry for each unit is located in a court area that is open to the adjacent unit. This court is approximately 6 feet by 12 feet. The 3-foot front entry doors are noted on the plan. Our department has four questions related to this drawing. First, we know the fire separation distance is based upon an imaginary lot line, or real lot line (such as in this drawing) running between the dwelling units and extending out to a public way, including the open court on this plan. Can this open court exist between Unit A and Unit B without a continuation of the 2 hours of fire-resistance-rated wall assemblies extending between the units through the open court?

A: Unit A and Unit B can remain open within the court, as long as the fire-resistance-rated separa-

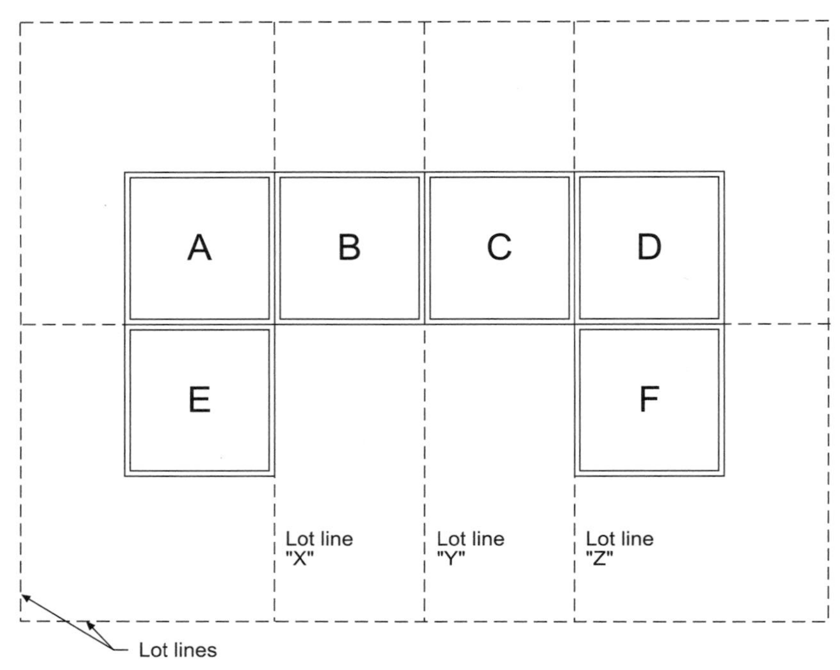

OPEN SPACE FOR SIX-UNIT TOWNHOUSE
FIGURE 3-58

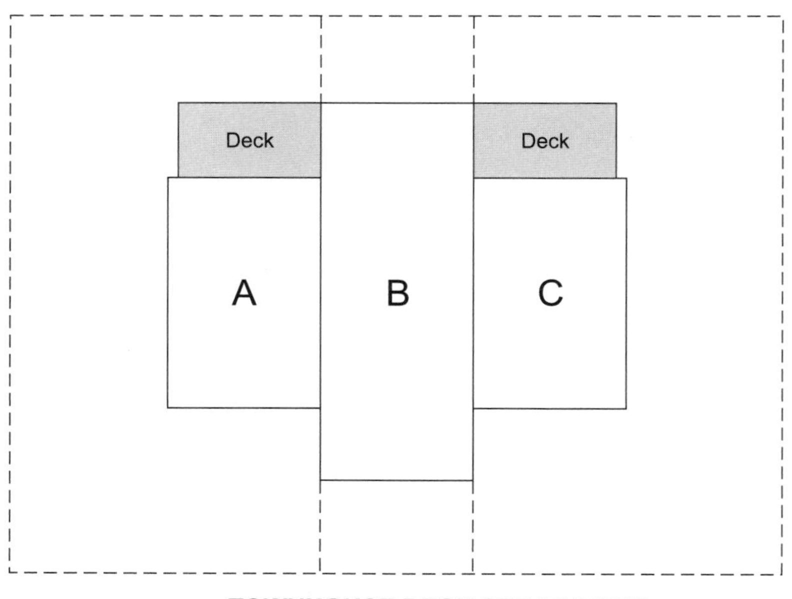

TOWNHOUSE DECK SEPARATION
FIGURE 3-59

tion extends though the attic above the court and out to the fascia line of the roof overhang above the court.

Q: Our second question, based on the same drawing, is related to Wall Y of dwelling Unit A. This wall is 6 feet from the lot line between Unit A and Unit B. Is it required to be a fire-resistance-rated wall assembly?

A: No. It meets the fire separation distance requirements of Section R302.

Q: Third, are window openings permitted in the same wall?

A: Yes. It meets the fire separation distance requirements of Section R302, which only places restrictions on window openings to 25 percent of the wall area if the wall is between 3 feet and 5 feet of the lot line.

Q: Fourth, the front entry doors are located in a wall that is perpendicular to the fire-resistance-rated walls that run along the lot line front to back. Are these front entry doors or the wall that they are located in required to be fire-resistance rated?

A: No. The provisions in Section R302.1 do not apply to walls perpendicular to the line used to determine the fire separation distance.

Q: When a townhouse is constructed with tuck-under garages as shown in Figure 3-61, do the requirements for fire separation between the attached (tuck-under) garage and the dwelling unit change because they are contained within a townhouse?

A: No. The separation between the dwelling unit and the tuck-under garage will need to meet the minimum requirements of Section R309, and any additional requirements based on location (fire separation distance).

Q: In this example, there are sleeping rooms located above the garage. We understand the ceiling of the garage will need to be protected with one layer of $^{5}/_{8}$-inch Type X gypsum board properly applied, but would the one layer of $^{1}/_{2}$-inch regular gypsum board on the walls below this ceiling meet the minimum requirements of Section R309.2.

A: Section R317.2 considers each townhouse to be a separate building. This section would require either two 1-hour fire-resistance-rated wall assemblies

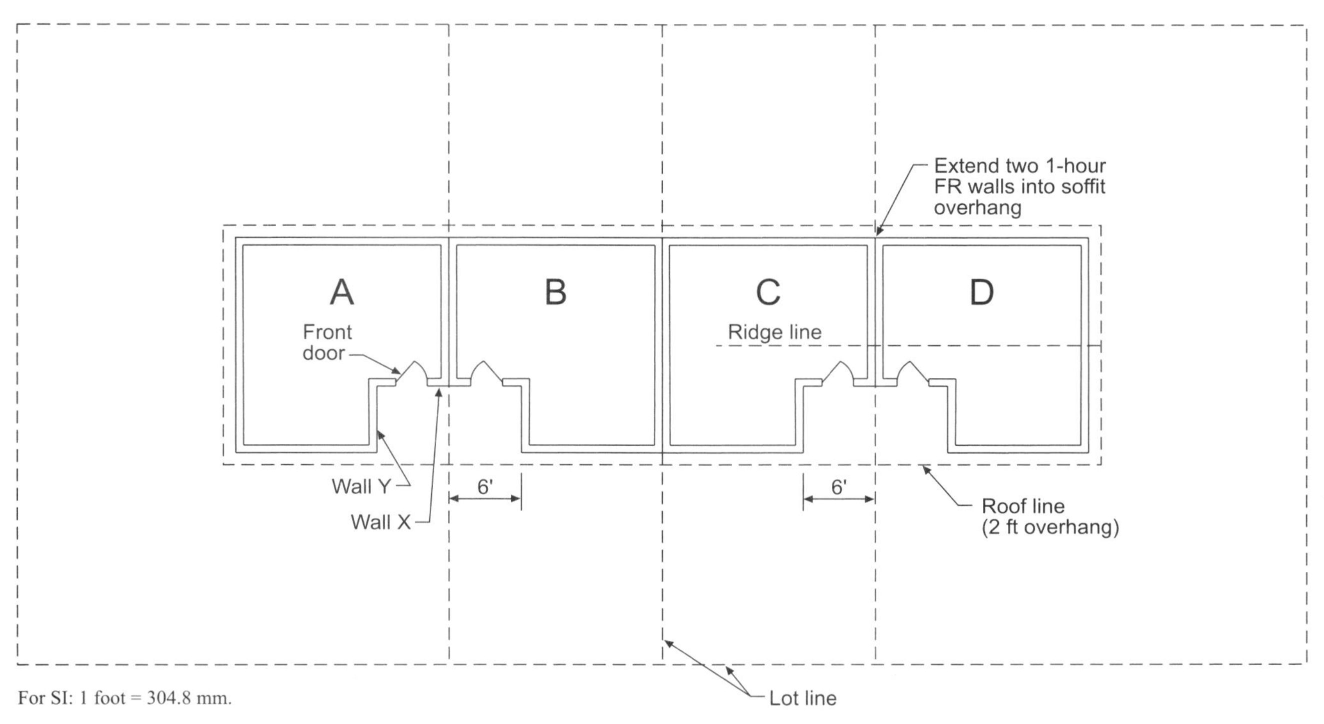

FOUR-UNIT TOWNHOUSE WITH FRONT FOYER AREAS
FIGURE 3-60

or one 2-hour fire-resistance-rated wall assembly between the garages. The ½-inch gypsum board can only be used if it is part of an approved assembly providing the required fire-resistance rating.

TOWNHOUSE WITH TUCK-UNDER GARAGES
FIGURE 3-61

Q: We have three general questions regarding exiting in townhouses. First, are there any provisions in Section R317.2 for townhouses, or elsewhere in the code, that would require more than one exit from a dwelling within a townhouse project based on the amount of stories or habitable space on the third level?

A: Dwelling units in the IRC only require one exit. There are no provisions in the code that require additional exits based on occupant load, number of stories or travel distance.

Q: Does the code limit the travel distance to the required exit?

A: No. Travel distance is not limited.

Q: Are there any provisions in the IRC that would require that a townhouse be constructed with 1-hour fire-resistant construction throughout?

A: No. The area of an individual dwelling unit or that of a dwelling unit in a group of three or more dwelling units, such as a townhouse, are not limited in area based on the type of construction.

R317.2.2 Parapets. Parapets constructed in accordance with Section R317.2.3 shall be constructed for townhouses as an extension of exterior walls or common walls in accordance with the following:

1. Where roof surfaces adjacent to the wall or walls are at the same elevation, the parapet shall extend not less than 30 inches (762 mm) above the roof surfaces.
2. Where roof surfaces adjacent to the wall or walls are at different elevations and the higher roof is not more than 30 inches (762 mm) above the lower roof, the parapet shall extend not less than 30 inches (762 mm)above the lower roof surface.

   Exception: A parapet is not required in the two cases above when the roof is covered with a minimum class C roof covering, and the roof decking or sheathing is of noncombustible materials or approved fire-retardant-treated wood for a distance of 4 feet (1219 mm) on each side of the wall or walls, or one layer of $^5/_8$-inch (15.9 mm) Type X gypsum board is installed directly beneath the roof decking or sheathing, supported by a minimum of nominal 2-inch (51 mm) ledgers attached to the sides of the roof framing members, for a minimum distance of 4 feet (1220 mm) on each side of the wall or walls.

3. A parapet is not required where roof surfaces adjacent to the wall or walls are at different elevations and the higher roof is more than 30 inches (762 mm) above the lower roof. The common wall construction from the lower roof to the underside of the higher roof deck shall have not less than a 1-hour fire-resistence rating. The wall shall be rated for exposure from both sides.

Q: Figure 3-62 shows a townhouse constructed with three dwelling units in a row. Does the two 1-hour or one 2-hour fire-resistance rated common wall (or walls) need to extend above the roof sheathing?

A: Yes. Parapets are to be constructed per Section R317.2.2.

Q: How high above the roof sheathing does the parapet need to extend?

A: If the roof surfaces of adjacent dwelling units are at the same elevation, the parapet needs to extend at least 30 inches above the roof.

Q: If a townhouse is constructed where every other dwelling unit has a second story that extends approximately 4 feet above the adjacent dwelling unit (see Figure 3-63), does the exterior wall of the second story of the adjacent dwelling unit need to be a 2-hour fire-resistance-rated assembly?

A: No. In the photograph, the one-story dwelling unit would require a 1-hour fire-resistance-rated assembly up to its roof sheathing. The adjacent two-story dwelling unit would require a 1-hour fire-resistance-rated assembly up to its roof sheathing.

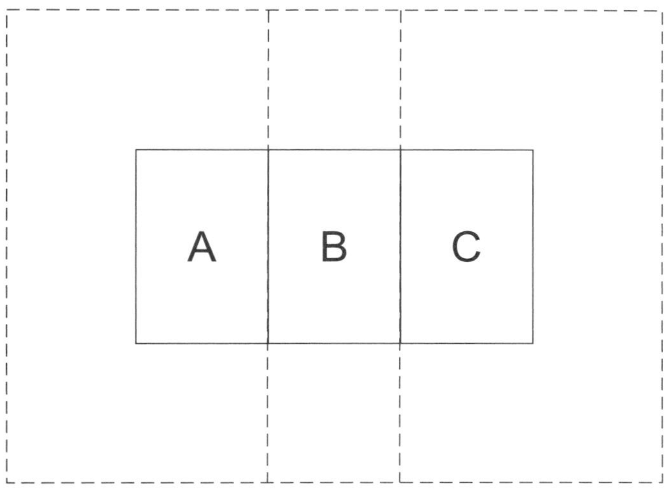

THREE-UNIT TOWNHOUSE
FIGURE 3-62

CHANGE IN ROOFLINE AT ADJACENT UNITS IN A TOWNHOUSE
FIGURE 3-63

Q: To follow-up the previous question, would a parapet need to be installed higher than the roof sheathing of the second story dwelling unit in the photograph?

A: No. Because these dwelling units are separated by more than 30 inches of elevation, no parapet is required.

Q: Could a window be installed in the gable end wall of the two-story dwelling unit overlooking the lower dwelling unit?

A: No. Openings are not permitted in the exterior wall elements with a fire separation distance less than 3 feet per Table R302.1.

Q: Could a window with fire-rated glass be installed in the gable end of the two-story dwelling unit overlooking the lower dwelling unit?

A: No. The same table applies.

Q: Are there any exceptions to the parapet requirement?

A: Yes. Also contained in Section R317.2.2, there is a parapet exception that allows the use of gypsum board or fire-retardant-treated wood for a distance of 4 feet on each side of the wall or walls.

Q: If a townhouse was constructed with fire-retardant-treated wood (roof sheathing) or gypsum board (below the non-fire-retardant-treated wood roof sheathing), are openings permitted in the sheathing within this 4 feet on each side of the common wall? The common wall between the units is located where the vertical brick line meets the adjacent dwelling unit.

A: Yes. There are no provisions that regulate the openings in the roof. This would permit roof vents, skylights, plastic vent stacks and other penetrations in the roof.

Q: Figure 3-64 shows the construction of a townhouse project. The fire-retardant-treated plywood sheathing is being installed 4 feet on each side of the fire-resistant separation walls below. Are the 4-foot by 8-foot sheets required to be laid in a staggered pattern?

A: No. There are no requirements in the code that require roof sheathing to be laid in a staggered pattern.

FIRE-RETARDANT-TREATED PLYWOOD ROOF SHEATHING ON TOWNHOUSE
FIGURE 3-64

Q: Our department has three questions related to the townhouse in Figure 3-65. First, in the front wall there are some windows, a main entry door and the overhead garage door. How close to the common wall separating the dwelling units can these unprotected window and door openings be located without requiring any fire-resistance rating?

A: Refer to Section R302.1. The provisions for the fire-resistance-rated wall assemblies between the dwelling units "shall not apply to walls, projections, openings or penetrations in walls that are perpendicular to the line used to determine the fire separation distance." The front windows and door openings in the photo can be unprotected, and there is no minimum distance required from the fire-resistance-rated wall assemblies between the dwelling units.

Q: Second, is the 2 feet of roof overhang considered a "projection" for the purposes of fire separation?

A: No. The 2 feet of roof overhang (or projection) is part of the wall that is perpendicular to the wall used to determine the fire separation distance. The 2-hour fire-resistance-rated assembly (two 1-hour or one 2-hour wall) needs to extend between the two dwelling units and extend into the soffit area in-line with the fire separation distance line.

WINDOW OPENINGS AT FRONT WALL OF TOWNHOUSE
FIGURE 3-65

Q: Third, soffit vents are installed in the 2 feet of roof overhang. Are the soffit vents permitted within 4 feet of the fire-resistance-rated wall assemblies (common wall)?

A: Yes. There are no requirements for the soffit to be protected.

Q: In a three-story dwelling unit of a townhouse, are shafts required for mechanical ductwork extending from the lowest level to the roof?

A: No. The IRC does not contain shaft enclosure requirements for townhouses. Section R602.8 requires fireblocking to cut off all concealed draft openings (both vertical and horizontal) and to form an effective fire barrier between stories and between a top story and the roof space.

Q: Does the IRC permit distribution lines for such utilities as gas and electrical lines as well as water and waste-water distribution lines to pass through the two 1-hour fire-resistant-rated common walls between the townhouse dwelling units?

A: Each townhouse shall be considered a separate building per Section R317.2. Other provisions of the code address these utilities:

- Section G2515.1 (404.1) Prohibited locations. Piping installed downstream of the point of delivery shall not extend through any townhouse unit other than the unit served by such piping.
- Section E3501.3 One building or other structure not to be supplied through another. Service conductors supplying a building or other structure shall not pass through the interior of another building or other structure.
- E3501.6.2 Service disconnect location. Each occupant shall have access to the disconnect serving the dwelling unit in which they reside.

R101.2 Scope. The provisions of the *International Residential Code for One- and Two-family Dwellings* shall apply to the construction, alteration, movement, enlargement, replacement, repair, equipment, use and occupancy, location, removal and demolition of detached one- and two-family dwellings and townhouses not more than three stories above-grade in height with a separate means of egress and their accessory structures.

Q: Section R101.2 limits a townhouse to not more than three stories above grade in height. For the purposes of fire and life safety, does the IRC have a limitation of building height in feet?

A: No. The IRC does not set a limit for building height in feet. The prescriptive provisions of Chapter 6, "Wall Construction," may effectively limit height depending on the method of construction. For example, wood and cold-formed steel studs are limited in length according to the applicable tables unless they are part of an engineered design.

R311.4 Doors.

R311.4.1 Exit door required. Not less than one exit door conforming to this section shall be provided for each dwelling unit. The required exit door shall provide for direct access from the habitable portions of the dwelling to the exterior without requiring travel through a garage. Access to habitable levels not having an exit in accordance with this section shall be by a ramp in accordance with Section R311.6 or a stairway in accordance with Section R311.5.

R311.4.2 Door type and size. The required exit door shall be a side-hinged door not less than 3 feet (914 mm) in width and 6 feet 8 inches (2032 mm) in height. Other doors shall not be required to comply with these minimum dimensions.

Q: If each of the dwelling units in a townhouse is considered a separate building, does the code require more than one exit based on height, stories, travel distance or occupancy load?

A: No. Only one exit is required.

SECTION R318 MOISTURE VAPOR RETARDERS

R318.1 Moisture control. In all framed walls, floors and roof/ceilings comprising elements of the building thermal envelope, a vapor retarder shall be installed on the warm-in-winter side of the insulation.

VAPOR RETARDER. A vapor resistant material, membrane or covering such as foil, plastic sheeting, or insulation facing having a permeance rating of 1 perm ($5.7 \cdot 10^{-11}$ kg/Pa $\cdot$ s $\cdot$ m^2) or less, when tested in accordance with the dessicant method using Procedure A of ASTM E 96. Vapor retarders limit the amount of moisture vapor that passes through a material or wall assembly.

Q: Is a 4 mil poly considered a vapor retarder?

A: Verify the particular specifications of the poly for the permeability rating, which must be 1 perm or less.

Q: Does spray-applied foam plastic insulation need to have a vapor retarder on the warm side?

A: Once again, it depends on the permeability rating. It you take the perm rating per inch, and divide it by the thickness of the foam insulation, it will give you the overall perm rating. For instance, if the foam had a perm rating of 1.75 perm per inch, and it was sprayed 3 inches thick, the overall perm rating would be about 0.58, and it would qualify as a vapor retarder by itself.

SECTION R319 PROTECTION AGAINST DECAY

R319.1 Location required. Protection from decay shall be provided in the following locations by the use of naturally durable wood or wood that is preservative treated in accordance with AWPA U1 for the species, product, preservative and end use. Preservatives shall be listed in Section 4 of AWPA U1.

1. Wood joists or the bottom of a wood structural floor when closer than 18 inches (457 mm) or wood girders when closer than 12 inches (305 mm) to the exposed ground in crawl spaces or unexcavated area located within the periphery of the building foundation.
2. All wood framing members that rest on concrete or masonry exterior foundation walls and are less than 8 inches (203 mm) from the exposed ground.
3. Sills and sleepers on a concrete or masonry slab that is in direct contact with the ground unless separated from such slab by an impervious moisture barrier.
4. The ends of wood girders entering exterior masonry or concrete walls having clearances of less than 0.5 inch (12.7 mm) on tops, sides and ends.
5. Wood siding, sheathing and wall framing on the exterior of a building having a clearance of less than 6 inches (152 mm) from the ground.
6. Wood structural members supporting moisture-permeable floors or roofs that are exposed to the weather, such as concrete or masonry slabs, unless separated from such floors or roofs by an impervious moisture barrier.
7. Wood furring strips or other wood framing members attached directly to the interior of exterior masonry walls or concrete walls below grade except where an approved vapor retarder is applied between the wall and the furring strips or framing members.

Q: What is pressure-preservative-treated wood?

A: Wood impregnated under pressure with chemical compounds that reduce the wood's susceptibility to deterioration caused by fungi, insects and marine borers.

Q: How is this wood treated?

A: The lumber is placed in a large horizontal cylinder. The door is sealed, and a vacuum is applied to remove most of the air from the cylinder and the wood cells. The preservative solution is pumped into the cylinder and the pressure is raised to a predetermined level, forcing the preservative solution into the wood. See Figure 3-66.

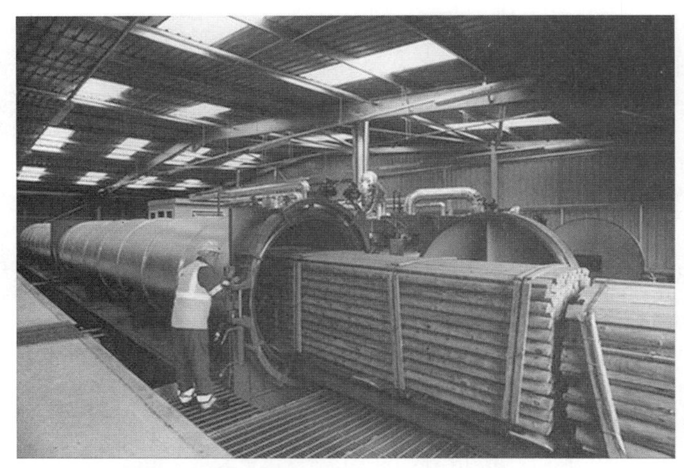

UNTREATED WOOD PLACED IN CYLINDER PRIOR TO TREATMENT
FIGURE 3-66

Q: Is it dried to any certain level after treatment?

A: Most treated wood is not dried after treatment. When it is dried after treatment, it is required to be dried to a moisture content of not more than 19 percent for lumber and 15 percent for plywood.

Q: Is the wood required to be kiln dried after treatment?

A: No. Kiln dried after treatment, commonly known in the industry as KDAT, is optional, except it is required for foundation grade-treated lumber and plywood.

Q: Are the wood treating plants monitored?

A: The American Lumber Standards Committee (ALSC) Treated Wood Program, operating under the Board of Review, has accredited independent third-party agencies in the United States and Canada. The operations of these agencies extend to over 230 treating plants operating in the two countries. Monitoring of treating plant production takes place under standards written and maintained by the American Wood Preservers Association (AWPA) and ALSC policies.

Q: What kind of items are checked during these plant inspections?

A: Items such as preservative penetration, preservative retention, drying after treatment, heartwood restriction and record keeping are checked during plant inspections.

Q: Are these chemical treatments regulated?

A: All pressure-preservative-treated wood chemicals are regulated by the United States Environmental Protection Agency (EPA) as pesticides. All pesticides sold or distributed in the United States must be registered by the EPA, based on scientific studies showing that they can be used without posing unreasonable risks to people or the environment.

Q: Are these pesticides reevaluated from time to time?

A: The law requires that pesticides that were first registered before November 1, 1984 be reregistered to ensure that they meet today's more stringent safety standards.

Q: Our department had heard that chromated copper arsenate (CCA) cannot be used anymore in residential construction. Is this true?

A: No. On February 12, 2002, the EPA announced a voluntary agreement with the industry to move consumer use of treated wood products away from a variety of pressure-treated wood that contains arsenic by December 31, 2003 in favor of alternative wood preservatives. CCA may still be used for permanent wood foundation, plywood, highway construction, round fence posts, marine construction, lumber and timber for salt water use and some others. This has resulted in very little CCA-treated wood being available or appropriate for residential use.

Q: The first paragraph of Section R319.1 refers to AWPA U-1 for the use and installation of pressure-preservative-treated wood. How was that developed?

A: AWPA U-1 and other AWPA standards are developed by its technical committees in an open, ANSI-approved consensus-based process that involves producers of preservatives, producers of treated wood products, end users of treated wood, engineers, architects, building code officials and other interested parties with a general interest in wood preservation.

Q: Are the AWPA standards voluntary or required?

A: The AWPA standards are voluntary standards, unless specifically referenced in a code, such as Section R319.1.

Q: We believe the retention of preservative in the wood based on the standards has changed in the last few years due to the changes with different types of treatments now used. Is this true?

A: No. The standards now include new preservatives, some which have different retention requirements than those that a user may be accustomed to. For example, if the user of a particular treated wood chemical, such as CCA, was normally using 0.40 pcf CCA for a ground contact deck post, and he or she is now using a different type of treated wood for the same application, it is possible that the wood he or she is now using has close to half the amount of treatment than was needed for the CCA. As long as any treatment is listed for a particular AWPA use category designation, it is appropriate for use regardless of the retention.

Q: What is a use category designation?

A: The Commodity Standards that have been used by the AWPA for years have been phased out for the new Use Category System (UCS) designation. This is used to designate what preservative system and retention level is needed to be effective in protecting wood products under specified exposure conditions.

Q: How many categories are there in this system?

A: There are five major categories that are broken down into subcategories.

Q: Based upon this AWPA table, titled "Service Conditions for Use Category Selection Guide," if I were to build an exterior deck using pressure-preservative-treated wood, the joists would need to be Use Category UC3B, and the deck posts in contact with the ground would need to be at least Use Category UC4A?

A: Correct.

Q: When a dwelling is constructed with a wood foundation, would the use category need to be at least UC4B?

A: Yes.

Q: In UC4A it notes ground contact and "critical components or difficult replacement." Would a deck post in contact with the ground be in a different use category than a porch post in contact with the ground?

A: Yes. The deck post as mentioned earlier would be a UC4A, where the porch post would need to be a UC4B because of the additional structural consideration of carrying a roof above; thus, this would qualify as a critical component or difficult replacement.

Q: Are these use categories required to be indicated on a grade stamp on the lumber?

A: They are required to be on an end tag or grade stamp per the AWPA and Section R319.2.1.

Q: We have heard a lot about CCA, but what other chemicals are being used for pressure-preservative-treated wood?

A: Section 4 of the referenced document AWPA U1 contains a listing of all of the chemical treatments.

Q: What is naturally durable wood?

A: As noted in Section R202, naturally durable wood is "the heartwood of the following species: Decay-resistant redwood, cedars, black locust and black walnut."

Q: Can both pressure-preservative-treated wood and wood that is naturally durable be used for all of the seven locations listed in Section R319.1?

A: Yes.

Q: Why would we use one over another?

A: Appearance, cost and availability are considerations. Also, there are provisions further in the section that will not permit the use of naturally durable wood, such as when it is buried in the ground or encased in concrete.

Q: If the wood joists noted in Item 1 of Section R319.1 were raised to 19 inches above the exposed earth, but were still within the periphery of the building foundation, then we understand that those wood joists would not need to be treated or naturally durable wood?

A: Correct. See Figure 3-67.

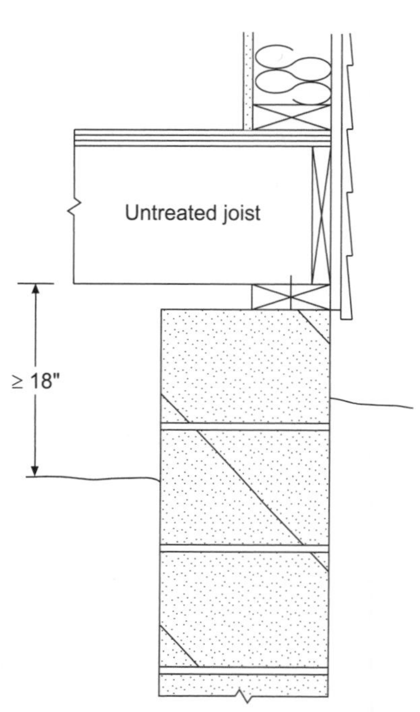

WOOD JOISTS ABOVE CRAWL SPACE
FIGURE 3-67

Q: To follow up our previous question, if the wood joists were 17 inches from the exposed ground in the crawl space within the periphery of the foundation, and an approved poly vapor barrier was laid on top of the exposed ground, would those wood joists still need to be treated or naturally durable wood?

A: No.

Q: In Item 2, we believe that if the top of the foundation is greater than 8 inches above the exposed ground, then the foundation sill plate would not need to be treated or a naturally durable wood.

A: That is correct. See Figure 3-68.

SILL PLATE ON FOUNDATION WALL
FIGURE 3-68

Q: What species of wood could the sill plate be if the top of the foundation is greater than 8 inches above the exposed ground?

A: It could be of any species recognized by the code as long as it was of sufficient strength to carry the loads prescribed by the code.

Q: If we use an untreated wood or one that is not naturally durable for the sill plate resting on the foundation wall because it is going to be greater than 8 inches above the exposed earth, and just before the final inspection the landscaper raises the exterior grade above that level, is there a code violation?

A: Yes. The final grade would need to be lowered.

Q: An untreated sill plate was installed on the top of the masonry foundation wall because the original plan was to keep the grade greater than 8 inches below the top of the foundation. The new homeowners took care of the final grading and the laying of sod, but they raised the sod to within 2 inches of the top of the foundation. Instead of removing all the new sod and regrading, the homeowners asked if they could remove the lower section of siding and install some treated plywood or metal flashing that would extend up and over the lower 12 inches of untreated wood plate and wall studs. Would this meet the intent of the code?

A: No. The entire exterior wall would need to meet the requirements as a wood foundation, Use Category UC4B, in order to comply.

Q: A stud wall is constructed on top of a concrete slab in a basement. Item 3 would require that the bottom sill plate be treated or naturally durable wood unless separated from the slab by an impervious moisture barrier. Could an untreated fir stud grade 2 x 4 plate be laid in contact with this concrete floor if a 6 mil poly vapor barrier was installed below the concrete floor prior to pouring it?

A: Yes. Protection against decay is only required for the plate when the slab is in direct contact with the ground.

Q: What if the project is a renovation in an existing house, and you are not sure if there is poly under the concrete slab. Could you install a layer of Type I #15 felt paper on the concrete floor to meet this requirement and then use the untreated fir plate?

A: If the building official determined that this Type I felt was an impervious moisture barrier as required by Section R319.1, it could be approved.

Q: If an untreated wood beam was installed in a beam pocket, and there was ½-inch air space on the top, sides and end, we understand that it would not need to be treated or be naturally durable wood. Would this beam need to have any impervious membrane separating the beam from the concrete or masonry foundation wall?

A: No. This wood beam could rest directly on the exposed concrete or masonry.

Q: We understand that wood siding needs to be kept at least 6 inches above the exposed ground, but could aluminum or vinyl siding be dropped down to an inch or so above the ground

A: Yes, the code would allow it. Certainly verify any special requirements with the manufacturer of the siding and keep in mind that any wood sheathing and framing still needs to maintain the clearance required.

Q: 2 x 4 studs were installed in a new basement with a layer of poly between the wood studs and the concrete foundation wall. If the studs were kept ¼ inch away from the concrete foundation wall, would the poly still be required by Item 7 of Section R319.1?

A: No.

R319.1.1 Field treatment. Field-cut ends, notches and drilled holes of preservative-treated wood shall be treated in the field in accordance with AWPA M4.

Q: What kind of treatment is required by AWPA M4 for treating the cut ends, notches and drilled holes out in the field?

A: The AWPA recognizes that many of the preservative systems are not packaged and labeled for use by the general public, and that a different product may need to be used for field treatment. Check with the manufacturer's representative to see what other products would be appropriate for that particular wood preservative. AWPA M4 recognizes copper naphthenate-based preservatives containing 2-percent copper for field treatment, and a number of products containing this preservative are available in the retail market. See Figure 3-69.

TREATED WOOD USED FOR STRUCTURAL MEMBERS OF DECK
FIGURE 3-69

R319.1.2 Ground contact. All wood in contact with the ground, embedded in concrete in direct contact with the ground or embedded in concrete exposed to the weather that supports permanent structures intended for human occupancy shall be approved pressure-preservative-treated wood suitable for ground contact use, except untreated wood may be used where entirely below groundwater level or continuously submerged in fresh water.

Q: Can the naturally durable woods be in contact with the ground or embedded in concrete in direct contact with the ground?

A: No. The wood would need to be pressure-preservative-treated wood with a use category designation of UC4A or greater. See Figure 3-70.

WOOD POST ON CONCRETE PIER
FIGURE 3-70

R319.1.5 Exposed glued-laminated timbers. The portions of glued-laminated timbers that form the structural supports of a building or other structure and are exposed to weather and not properly protected by a roof, eave or similar covering shall be pressure treated with preservative, or be manufactured from naturally durable or preservative-treated wood.

Q: We understand that a pressure-preservative-treated glue-laminated beam requires an end-tag or other grade mark that contains all the required information noted in Section R319.2.1. Does this apply when used as a structural support of an exterior deck?

A: Yes.

R319.2 Quality mark. Lumber and plywood required to be pressure-preservative-treated in accordance with Section R319.1 shall bear the quality mark of an approved inspection agency that maintains continuing supervision, testing and inspection over the quality of the product and that has been approved by an accreditation body that complies with the requirements of the American Lumber Standard Committee treated wood program.

Q: If a load of treated lumber arrives to a job site without the quality mark of an approved inspection agency noted on the end tag, yet all other required information is noted on the end tag, can the building department reject the lumber?

A: Yes. The name or logo of the inspection agency is required to be noted on the tag. See Figure 3-71.

END TAG ON TREATED WOOD POST
FIGURE 3-71

Q: Some of the local lumber suppliers are selling solid sawn hardwoods from South America as exterior decking material. These 1 x 4 and 1 x 6 boards are not composite materials and there are no grade stamps on the wood. Would these hardwoods be recognized as boards that can be used for exterior decking by the code?

A: No. The definition of "Naturally durable woods" only includes redwood, cedar, black locust and black walnut.

Q: The supplier of some solid sawn hardwoods has provided engineering data for the particular wood, and would like us to approve it for use as deck boards based on the engineering. Would this information be sufficient?

A: There are a few items to consider. First, the lumber is not grade stamped. Even if it was grade stamped, the grade stamps for hardwoods are based on how clear the lumber is and not on structural considerations. If engineering was provided for a particular exotic hardwood that appeared to meet minimum structural requirements, and there was no grade stamp, the field inspector would not know what kind of wood it was, or even if it was the same wood that the engineering was provided for. Inspectors are not usually certified as lumber evaluators; they are taught to look for proper grade stamps.

Q: We understand that borate-treated wood can only be used for interior use. We have noticed lately that some borate-treated wood is being promoted for exterior use due to some special sealing of the wood. Is this possible?

A: Borate-treated wood, most commonly treated with disodium octaborate tetrahydrate, as noted in the ICC ES Report ER-4890 and AWPA standards, is limited to "locations that are not in contact with the ground and are not subject to contact with liquid water." This includes AWPA Use Category UC1 and UC2. There are currently no borate-treated woods with sealers listed with the AWPA or that possess an ICC ES report that would recognize it for exterior use.

Q: We have a builder in town who has chosen to use 4 x 4 treated posts laid flat instead of a concrete curb block on top of the interior wall footing in his basements. The wood is tagged for "Ground Contact." Would this comply with the code?

A: This pressure-preservative-treated wood is correctly tagged for the use. You should also verify the method of installation related to anchoring of this material.

R319.3 Fasteners. Fasteners for pressure-preservative and fire-retardant-treated wood shall be of hot-dipped zinc-coated galvanized steel, stainless steel, silicon bronze or copper. The coating weights for zinc-coated fasteners shall be in accordance with ASTM A 153.

Exceptions:

1. One-half-inch (12.7 mm) diameter or larger steel bolts.
2. Fasteners other than nails and timber rivets shall be permitted to be of mechanically deposited zinc-coated steel with coating weights in accordance with ASTM B 695, Class 55, minimum.

Q: Does ASTM A 153 require a minimum amount of zinc on a fastener?

A: Yes. Fasteners that meet ASTM A 153, *Standard Specification for Zinc Coating by the Hot Dip Process,* are coated with 1 ounce of zinc coating per square foot.

Q: Based on Section R319.3, all of the electroplated toenails that are being used to fasten down the floor joists or floor trusses to a pressure-preservative-treated sill plate do not meet code. Is this correct?

A: Yes, that is correct. Electroplated nails have roughly $^1/_{10}$ the zinc coating as hot-dipped galvanized nails meeting ASTM A 153. There are also significant differences in the application process.

Q: Would this include the 8d nails that are being used to nail the exterior wall sheathing to the sill plate?

A: Yes, if they are also nailed into the pressure-preservative-treated sill plate.

Q: Does the word "fasteners" refer to nails, screws, bolts and joist hangers?

A: Fasteners are nails, screws and bolts, but not joist hangers. Joist hangers are considered connectors or hardware.

Q: There are some ICC ES reports by some of the major producers of some of the pressure-preservative chemical treatments that state in their reports: "Metals used in contact with (this preservative) pressure treated wood shall be hot dip galvanized, stainless steel or triple coated zinc polymer materials. Carbon steel, aluminum, red brass and bronze shall not be used in contact with (these preservative) treated wood products." Would this apply to all metals in contact with the preservative, such as flashing material, nails, bolts, etc?

A: It would apply to all metals unless specifically exempted from the requirement, such as Exception 1 for ½-inch-diameter or larger steel bolts. See Figure 3-72.

Q: Why are the bolts exempt if a pressure-preservative-treated foundation sill plate is installed that may be in contact with the chemical treatment?

A: Although the manufacturers of the chemical treatments have a general statement about the use of the wood in contact with metal, the ½-inch-diameter carbon steel anchor bolts are exempt for two reasons. First, the possible reaction between the wood treatment and metal is a result of a wet/dry cycle that would not normally occur in a protected area of a rim joist (or sill plate and header). Second, the mass of the ½-inch-diameter bolt would prevent the bolt from losing significant structural integrity due to any minor corrosion.

FOUNDATION ANCHOR BOLT
FIGURE 3-72

Q: Our department understands that connectors or hardware such as joist hangers and foundation anchor straps are not required to meet the ASTM A 153 standard that is required for fasteners. Are they required to meet any other particular standard related to the zinc-coating thickness?

A: The code does not require compliance with any referenced standard for the coating thickness.

Q: We have read some of the ICC ES reports for materials such as joist hangers and other connectors that indicate coating thicknesses of zinc based on ASTM A 924/A and A 653. Are these standards required by the code or the ICC ES?

A: No. They are included in the report because they are contained in the manufacturer's submittals.

Q: Is there an industry standard for zinc-coating thickness for joist hangers for exterior use?

A: The industry standard generally appears to be a G-60 coating thickness. That is 0.60 ounces of zinc per square foot of area.

Q: What affect, if any, do the alternative chemicals have on fasteners, hardware and connectors?

A: The industry is continually testing these chemical treatments related to their products. Some manufacturers have recommended the use of higher levels of zinc coating on their products based on their own testing.

Q: We have heard the term "G-185" recommended by one manufacturer of hardware. Does this mean it is recommending a coating thickness that is about three times the industry standard?

A: Yes.

Q: If the code does not set a minimum coating thickness for connectors and hardware as it does for fasteners (ASTM A 153), what should a building official look for when he or she is trying to approve an installation?

A: Some of the manufacturers of connectors and hardware are recommending coating thicknesses based on their own independent testing in order for their product to function as intended over the life of the product and to ensure a minimum level of safety for the user. Although the code does not address it at this time, the building official may want to use the most current data available for that particular product.

NOTES

FOUNDATIONS

CHAPTER

4

SECTION R401 GENERAL

R401.1 Application. The provisions of this chapter shall control the design and construction of the foundation and foundation spaces for all buildings. In addition to the provisions of this chapter, the design and construction of foundations in areas prone to flooding as established by Table R301.2(1) shall meet the provisions of Section R324. Wood foundations shall be designed and installed in accordance with AF&PA Report No. 7.

Q: When designing a wood foundation according to the provisions in the code, would a registered design professional need to sign the plans?

A: No, unless required by the building official because of special conditions (Section R106.1) or the structural elements exceed the limits of the IRC prescriptive provisions (Section R301.1.3).

Q: Can a wood foundation support brick veneer?

A: Yes. Section R703.7.2 permits exterior stone and masonry veneer to be supported by wood. The design of the wood shall consider the weight of the veneer and any other load. The *Permanent Wood Foundation Guide to Design and Construction* also contains provisions for brick veneer. See Figure 4-1.

Q: Where are the provisions for cold weather masonry construction located in the code?

A: Section R607.1 makes reference to ASTM C 270, where the guidelines are located. Also, see Q & A for Section R607.1 in Chapter 6 and Figure 4-2.

Q: What precautions should be taken during cold weather for poured concrete footings and walls?

A: According to the American Concrete Institute (ACI), poured concrete shall be maintained above 50°F and in a moist condition for at least the first seven days after placement. ACI also has recommendations for cold weather concreting in a document titled "Cold Weather Concreting," reported by ACI Committee 306.

WOOD FOUNDATION WITH BRICK LEDGE
FIGURE 4-1

R401.2 Requirements. Foundation construction shall be capable of accommodating all loads according to Section R301 and of transmitting the resulting loads to the supporting soil. Fill soils that support footings and foundations shall be designed, installed and tested in accordance with accepted engineering practice. Gravel fill used as footings for wood and precast concrete foundations shall comply with Section R403.

Q: Could the building official require tests of soil conditions or concrete footings after they are installed to verify if they comply with the code?

A: The building official has the authority to require testing but also the responsibility to act reasonably, fairly and consistently. Only under extraordinary circumstances where the building official has justifiable and defensible cause should he or she take such action. Any questions related to soil properties should be resolved at the plan review stage. If there are concerns based on local conditions, experience or test data, the building official has authority to require a soils test prior to construction. If unanticipated adverse soil conditions are discovered during excavation, the builder should notify the building official. At the last check of soil conditions, the inspector is responsible for approving the base for footings before they are poured. Once inspection is approved and the footings poured, the building official must justify any action to withdraw the approval and cause hardship to the builder. The same responsibilities apply to the question of testing of the concrete after it is poured. The building official must first determine there are significant reasons for requiring testing.

COLD WEATHER PROTECTION FOR MASONRY CONSTRUCTION
FIGURE 4-2

Exception: The provisions of this chapter shall be permitted to be used for wood foundations only in the following situations:

1. In buildings that have no more than two floors and a roof.
2. When interior basement and foundation walls are constructed at intervals not exceeding 50 feet (15 240 mm).

Q: Could a wood foundation be used for a three-story dwelling?

A: It would be outside the prescriptive requirements of the code and, as such, would need to be designed by a licensed design professional and approved by the building official.

R401.3 Drainage. Surface drainage shall be diverted to a storm sewer conveyance or other approved point of collection so as to not create a hazard. Lots shall be graded to drain surface water away from foundation walls. The grade shall fall a minimum of 6 inches (152 mm) within the first 10 feet (3048 mm).

Exception: Where lot lines, walls, slopes or other physical barriers prohibit 6 inches (152 mm) of fall within 10 feet (3048 mm), the final grade shall slope away from the foundation at a minimum slope of 5 percent and the water shall be directed to drains or swales to ensure drainage away from the structure. Swales shall be sloped a minimum of 2 percent when located within 10 feet (3048 mm) of the building foundation. Impervious surfaces within 10 feet (3048 mm) of the building foundation shall be sloped a minimum of 2 percent away from the building.

Q: In Section R401.3, could the "other approved point of collection" for the surface drainage be a small pond area on the same lot as the dwelling?

A: Yes, with approval of the building official.

Q: When footings are designed using the amounts in Table R401.4.1, who determines what load-bearing pressure to use in the design.

TABLE R401.4.1
PRESUMPTIVE LOAD–BEARING VALUES OF FOUNDATION MATERIALS[a]

CLASS OF MATERIAL	LOAD-BEARING PRESSURE (pounds per square foot)
Crystalline bedrock	12,000
Sedimentary and foliated rock	4,000
Sandy gravel and/or gravel (GW and GP)	3,000
Sand, silty sand, clayey sand, silty gravel and clayey gravel (SW, SP, SM, SC, GM and GC)	2,000
Clay, sandy clay, silty clay, clayey silt, silt and sandy silt (CL, ML, MH and CH)	1,500[b]

For SI: 1 pound per square foot = 0.0479 kPa.

a. When soil tests are required by Section R401.4, the allowable bearing capacities of the soil shall be part of the recommendations.

b. Where the building official determines that in-place soils with an allowable bearing capacity of less than 1,500 psf are likely to be present at the site, the allowable bearing capacity shall be determined by a soils investigation.

A: In order to get started in the design of the structure, the designer will need to assume some load-bearing capability of the soil based upon his or her knowledge of the local building conditions and any other data provided. When the plan is submitted to the building department for review, any appropriate data should be included in the submittal documents. The building official will need to make a judgment call based on the submittal documents, but ultimately the field inspector will need to verify this at the time of the footing inspection. If the soil looks to be different than what was proposed with the submittal documents, the field inspector can require engineering analysis of the soil at that time.

Q: Is a soil report required as part of the submittal documents for a house permit?

A: The building official can require a soil report if soil conditions are unknown or suspected to be adverse.

SECTION R402 MATERIALS

R402.1 Wood foundations. Wood foundation systems shall be designed and installed in accordance with the provisions of this code.

R402.1.1 Fasteners. Fasteners used below grade to attach plywood to the exterior side of exterior basement or crawlspace wall studs, or fasteners used in knee wall construction, shall be of Type 304 or 316 stainless steel. Fasteners used above grade to attach plywood and all lumber-to-lumber fasteners except those used in knee wall construction shall be of Type 304 or 316 stainless steel, silicon bronze, copper, hot-dipped galvanized (zinc coated) steel nails, or hot-tumbled galvanized (zinc coated) steel nails. Electro- galvanized steel nails and galvanized (zinc coated) steel staples shall not be permitted.

R402.1.2 Wood treatment. All lumber and plywood shall be pressure-preservative treated and dried after treatment in accordance with AWPA U1 (Commodity Specification A, Use Category 4B and Section 5.2), and shall bear the label of an accredited agency. Where lumber and/or plywood is cut or drilled after treatment, the treated surface shall be field treated with copper naphthenate, the concentration of which shall contain a minimum of 2 percent copper metal, by repeated brushing, dipping or soaking until the wood absorbs no more preservative.

Q: What is the difference between Type 304 and Type 316 stainless steel?

A: Type 304 stainless steel is 18-percent chromium and 8-percent nickel. Type 316 is 16-percent chromium, 10-percent nickel and 2-percent molybdenum. The molybdenum is added to help resist corrosion to chlorides, such as sea water and de-icing salts. Both are appropriate for wood foundations. Type 304 is also used for kitchen appliances and vent hoods. Type 304 is not resistant to muratic acid. Grout cleaners should not be used around Type 304 stainless steel because the fumes will also cause discoloration of the stainless steel.

Q: Can chromate copper arsenate still be used for wood foundations?

A: Yes. It does need to meet the AWPA Use Category 4B standard. For CCA, that would work out to foundation grade (FDN) 0.60 treatment. When using other chemical treatments, refer to the end tag for the use category and to the manufacturer for other installation instructions.

SECTION R403 FOOTINGS

R403.1 General. All exterior walls shall be supported on continuous solid or fully grouted masonry or concrete footings, wood foundations, or other approved structural systems which shall be of sufficient design to accommodate all loads according to Section R301 and to transmit the resulting loads to the soil within the limitations as determined from the character of the soil. Footings shall be supported on undisturbed natural soils or engineered fill.

Q: A local contractor excavated approximately 2 feet deeper than the bottom of the exterior wall footings for a new house. He than filled the excavation with 2 feet of washed river rock, for drainage, as he noted. He then set his forms and called for a footing inspection. Our inspector asked for an engineer's analysis of the fill material prior to approval of the formwork. Does the code allow us to ask for this analysis?

A: Yes. Even though the material is washed river rock, it needs to be evaluated for compliance with the code.

Q: We have a situation where a townhouse project has a series of dwellings with attached garages.

Are the footings for the dwellings and the footings for the attached garages required to be poured continuous?

A: No. The code requires the footings to be continuous, but it does not require these concrete footings to be poured continuous. The code does not prohibit the use of construction joints in the footings. This design approach minimizes the effects of differential settlement since most houses are constructed without the benefit of an extensive soils investigation or a design engineer's recommendations. Continuous footings around the perimeter of a house and garage are necessary to safely transfer all of the loads acting on the structure down to the soils.

Q: When a garage is attached to a dwelling with frost depth footings around the perimeter, can the garage footings below the location of the overhead door be eliminated?

A: No. The footings below the overhead door are required, and they must also extend to the designated frost depth as the other garage wall footings.

> **R403.1.1 Minimum size.** Minimum sizes for concrete and masonry footings shall be as set forth in Table R403.1 and Figure R403.1(1). The footing width, W, shall be based on the load-bearing value of the soil in accordance with Table R401.4.1. Spread footings shall be at least 6 inches (152 mm) thick. Footing projections, *P*, shall be at least 2 inches (51 mm) and shall not exceed the thickness of the footing. The size of footings supporting piers and columns shall be based on the tributary load and allowable soil pressure in accordance with Table R401.4.1. Footings for wood foundations shall be in accordance with the details set forth in Section R403.2, and Figures R403.1(2) and R403.1(3).

Q: We typically install 20-inch-wide footings for our 12-inch-wide masonry walls, which provide at least 4 inches of footing projection on each side. What is the minimum projection required by the code?

A: Footing projections need to be at least 2 inches for spread wall footings. Many builders use the 4-inch standard to allow for any slight variation of the location of the footing to ensure at least 2-inch minimum projection. Also, the code requires that the footing projection not exceed the thickness of the footing to mitigate diagonal tension cracking. See Figure 4-3.

TABLE R403.1
MINIMUM WIDTH OF CONCRETE OR MASONRY FOOTINGS (inches)[a]

	LOAD-BEARING VALUE OF SOIL (psf)			
	1,500	2,000	3,000	≥4,000
Conventional light–frame construction				
1-story	12	12	12	12
2-story	15	12	12	12
3-story	23	17	12	12
4-inch brick veneer over light frame or 8-inch hollow concrete masonry				
1-story	12	12	12	12
2-story	21	16	12	12
3-story	32	24	16	12
8-inch solid or fully grouted masonry				
1-story	16	12	12	12
2-story	29	21	14	12
3-story	42	32	21	16

FOOTING PROJECTION BEYOND MASONRY WALL
FIGURE 4-3

Q: Our proposed structure is a three-story conventional light-framed dwelling with 4-inch brick veneer. The assumed soil bearing capacity is 2,000 psf. According to Table R403.1, the minimum footing width is 24 inches. Could the footing be less than 24 inches?

A: Structural calculations taking into account the actual conditions (actual wall heights, wall component weights, foundation wall height and weight, etc.)

could be submitted to the building official to substantiate a narrower footing width. If the building official approves these calculations as an alternative to the prescribed requirements, then a narrower footing width may be used.

Q: How are deck and porch footing sizes determined?

A: The designer of a deck or porch needs to first determine the live and dead loads of the floor and/or roof and wall systems, and what load will travel down to each post or pier footing. He or she would then need to determine the bearing capacity of the soil. For example, let us assume a 2,000 psf soil-bearing capacity. A post/pier has a design load coming down on it of 2,793 pounds. The load of 2,793 pounds is divided by the soil bearing capacity of 2,000 psf, which equals 1.39 square feet of area needed on the bottom of the footing. The designer wants to use a round footing. A circle 16 inches in diameter equals 1.40 square feet. Therefore, a concrete footing 16 inches in diameter will support the 2,793-pound design load.

R403.1.4 Minimum depth. All exterior footings shall be placed at least 12 inches (305 mm) below the undisturbed ground surface. Where applicable, the depth of footings shall also conform to Sections R403.1.4.1 through R403.1.4.2.

R403.1.4.1 Frost protection. Except where otherwise protected from frost, foundation walls, piers and other permanent supports of buildings and structures shall be protected from frost by one or more of the following methods:

1. Extended below the frost line specified in Table R301.2.(1);
2. Constructing in accordance with Section R403.3;
3. Constructing in accordance with ASCE 32; or
4. Erected on solid rock.

Exceptions:

1. Protection of freestanding accessory structures with an area of 600 square feet (56 m^2) or less, of light-framed construction, with an eave height of 10 feet (3048 mm) or less shall not be required.
2. Protection of freestanding accessory structures with an area of 400 square feet (37 m^2) or less, of other than light-framed construction, with an eave height of 10 feet (3048 mm) or less shall not be required.
3. Decks not supported by a dwelling need not be provided with footings that extend below the frost line.

Q: In Exception 1 it refers to "light-framed construction." What does this refer to?

A: In Section R202 it defines "Light-framed construction" as a type of construction with vertical and horizontal structural elements that are primarily formed by a system of repetitive wood or light gage steel framing members.

Q: When does the footing for a deck need to extend to the frost line?

A: When it is attached to the dwelling that also has footings that extend to the frost line.

Q: In our part of the country, the required frost depth is 42 inches below the finished grade level. A series of townhouses are being constructed with tuck-under garages and we have required the footings for the garages to extend to the 42-inch frost depth at all four sides. The builder believes that because these tuck-under garages are adjacent to conditioned spaces on at least two sides that the frost protection is not needed. Are we asking for too much?

A: No. The code requires frost depth footings at all sides of the garage. The building official could consider an engineered design indicating a lesser depth.

Q: What if the tuck-under garage was insulated and contained a gas unit heater?

A: No. The unit heater could be turned off by the homeowner, which in turn may allow the frost to travel down.

Q: If a deck was not physically attached to the dwelling, could it be placed on floating concrete pier blocks (plinth blocks)?

A: Yes. The wind load for the geographical area should also be considered to address any anchoring requirements that may need to be addressed.

Q: Are interior pad footings in a conditioned space, such as a heated basement, subject to the minimum frost depth provisions?

A: No. If these pad footings were in an unconditioned space, such as an attached garage, they would need to extend to the frost depth. See Figure 4-4.

PAD FOOTING IN CONDITIONED SPACE
FIGURE 4-4

Q: Does the code require the concrete basement floor to rest on the interior footing projection of the exterior wall footing?

A: No. The basement slab is a nonstructural component.

R403.1.6 Foundation anchorage. When braced wall panels are supported directly on continuous foundations, the wall wood sill plate or cold-formed steel bottom track shall be anchored to the foundation in accordance with this section.

The wood sole plate at exterior walls on monolithic slabs and wood sill plate shall be anchored to the foundation with anchor bolts spaced a maximum of 6 feet (1829 mm) on center. There shall be a minimum of two bolts per plate section with one bolt located not more than 12 inches (305 mm) or less than seven bolt diameters from each end of the plate section.

Q: There is a ½-inch-diameter anchor bolt in the top course of the foundation wall of an attached garage. The sill plate is a 2 x 4. How close to the edge of the sill plate can this anchor bolt be located?

A: Section R403.1.6 provides the requirements for size and spacing of the anchor bolts, but it does not refer to a minimum edge distance. For engineered wood-frame buildings, Table 8.5.3, "Edge Distance Requirements for Bolts," of the NDS published by the American Forest and Paper Association (AF&PA) provides a means to calculate the edge distance based on the size of the bolt, width of the sill plate and direction of loading to the grain.

Q: If that edge distance information is not provided in the body of the code can a building official enforce it?

A: No. The provisions in the code are prescriptive provisions. In Section R301.1, "Application," it generally states that a structure needs to be constructed in accordance with the provisions in the code. In Section R301.1.1, "Alternate provisions," it states, "The following standards are permitted subject to the limitations of the code ... Where engineering design is used in conjunction with these standards the design shall comply with the *International Building Code*." Then it goes on to reference the AF&PA *Wood Frame Construction Manual* (WFCM). The NDS is the engineering design standard for wood construction referenced in the IRC and IBC. The builder has the option of providing engineering data to substantiate the installation of that bolt location in relation to the edge of the sill plate in accordance with the NDS, or the builder can use the WFCM.

Q: Can the building official require compliance with the provisions in the NDS?

A: No. It is not the building official's option to require engineering unless the building or portions or elements of the building require engineering according to Section R301.1.3. The building official can require compliance with the provisions in the code, which may indeed require engineering.

Q: So, for this ½-inch-diameter anchor bolt in the 2 x 4 sill plate, with a wind load against the wall sheathing, could you tell us what the minimum edge distance is?

A: From an engineering perspective, for loads perpendicular to the grain, the minimum edge distance to the loaded edge is 4D (four times the diameter), or four times the ½-inch diameter of the bolt in this situation. The minimum edge distance would be 2 inches. Edge distance is measured from the edge of the member to the centerline of the nearest fastener.

Q: If the sill plate is actually 3½ inches wide, would this not comply?

A: Technically you are correct; the edge distance does not meet the engineering requirements, but the code does allow the 2 x 4 plate to be used, and the ½-inch-diameter bolt is the minimum for this situation.

Q: Does the code or NDS address the maximum diameter of the hole for the bolt?

A: Yes. The NDS specifies that bolt holes shall be a minimum of $^1/_{32}$ inch to a maximum of $^1/_{16}$ inch larger than the bolt diameter.

Q: The code requires a bolt within 12 inches of the end of each plate section. What options are there if this is not complied with at the time of the framing inspection?

A: The builder would need to provide engineering data to the building official for approval of a foundation anchor that could be installed after the plate is installed. This may be some sort of expansion-type bolt or similar approved product.

Q: In lieu of installing a foundation anchor in the plate, could the builder install a short section of 2 x 4 blocking over the spliced area of the plate to comply with this requirement?

A: The purpose of the anchor bolt is to hold the structure to the foundation. The 2 x 4 blocking would not provide any additional anchorage directly to the foundation wall.

R403.1.7.3 Foundation elevation. On graded sites, the top of any exterior foundation shall extend above the elevation of the street gutter at point of discharge or the inlet of an approved drainage device a minimum of 12 inches (305 mm) plus 2 percent. Alternate elevations are permitted subject to the approval of the building official, provided it can be demonstrated that required drainage to the point of discharge and away from the structure is provided at all locations on the site.

Q: If a house is located 50 feet from the street gutter, how much higher is the top of the foundation wall required to extend above the point of discharge?

A: The code requires a minimum of 12 inches plus 2 percent. In a distance of 50 feet, 2 percent is 12 inches. So, 12 inches plus 2 percent, or in this situation, 12 inches more, equals 24 inches between the top of the foundation and the street gutter at the point of discharge. See Figure 4-5.

SECTION R404
FOUNDATION AND RETAINING WALLS

R404.1 Concrete and masonry foundation walls. Concrete and masonry foundation walls shall be selected and constructed in accordance with the provisions of Section R404 or in accordance with ACI 318, ACI 332, NCMA

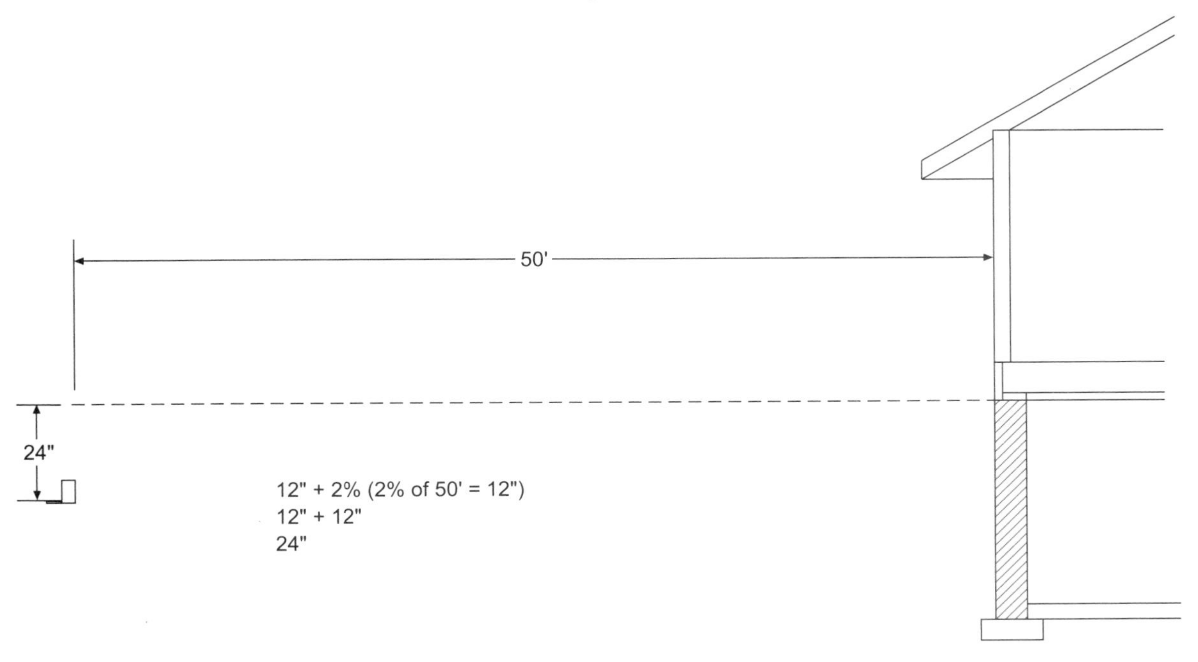

FOUNDATION ELEVATION ON GRADED SITE
FIGURE 4-5

TR68–A or ACI 530/ASCE 5/TMS 402 or other approved structural standards. When ACI 318, ACI 332 or ACI 530/ASCE 5/TMS 402 or the provisions of Section R404 are used to design concrete or masonry foundation walls, project drawings, typical details and specifications are not required to bear the seal of the architect or engineer responsible for design, unless otherwise required by the state law of the jurisdiction having authority.

Q: In Section R404.1 the code requires the foundation wall to be constructed according to the provisions in the code or in accordance with a small handful of other referenced documents. Why so many?

A: These documents, developed by the ACI, National Concrete Masonry Institute, American Society of Civil Engineers (ASCE) and The Masonry Society have been used by designers for many years. By specifically referencing these in the code, the building official and the designer will not have to use the provisions in the code for alternate materials to consider these options.

R404.1.3 Design required. Concrete or masonry foundation walls shall be designed in accordance with accepted engineering practice when either of the following conditions exists:

1. Walls are subject to hydrostatic pressure from groundwater.
2. Walls supporting more than 48 inches (1219 mm) of unbalanced backfill that do not have permanent lateral support at the top or bottom.

Q: In Item 2 of Section R404.1.3, "Design required," what type of wall is this section referring to?

A: It refers to a concrete or masonry foundation wall that does not extend from the footing up to the bottom of the floor joist system. The unbalanced fill height, that being the distance from the top of the basement floor to the height of the finished grade, exceeds 4 feet. The balance of the wall is normally framed as a knee wall or cripple wall, that being wall studs of sufficient size and spacing to carry the load. The issue here is that a pivot point (or hinge point) occurs where the top of the concrete or masonry wall meets the wood wall framing. This would also be known as a cantilevered wall. Because this wall cannot be laterally supported top and bottom due to the nature of this pivot point, it will require engineering design when the unbalanced backfill height exceeds 4 feet. See Figure 4-6.

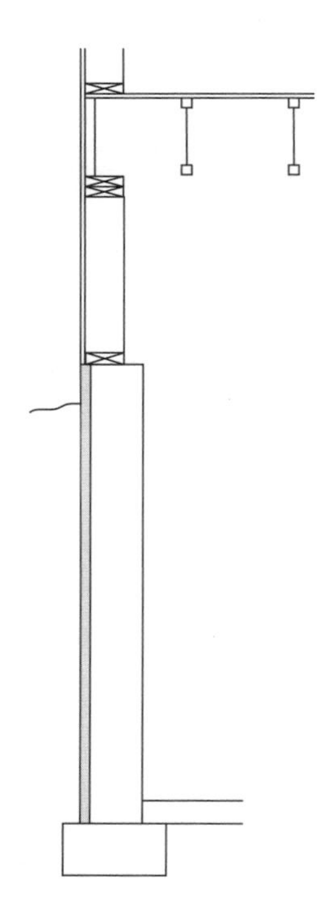

CONCRETE/MASONRY FOUNDATION WALL WITH CRIPPLE STUDS
FIGURE 4-6

Q: If a masonry foundation wall requires vertical reinforcing steel according to the tables in the code, besides grouting the reinforced cells, does the table require the entire wall to be filled with grout?

A: No. Only the cells where the reinforcing is located are required to be grouted.

Q: If vertical steel reinforcing is required in the walls, is it required to be doweled into the footings?

A: No. Section R404.1 requires foundation walls to be laterally supported. The foundation wall is considered laterally supported at the bottom when a 3 ½-inch-thick concrete slab is poured tight against the bottom of the wall. The code does not provide any prescriptive means to restrain the bottom of the foundation wall for basements without concrete slabs, and as such, the building official will most likely require engineering to verify the design.

Q: Note c of Table R404.1.1(4) (see below) refers to Grade 60 steel. What is Grade 60 steel?

A: Grade 60 steel reinforcing bars have a minimum yield strength of 60,000 psi. Refer to ASTM A 615, *Standard Specification for Deformed and Plain Bil-*

let-Steel Bars for Concrete Reinforcement, for additional information regarding the steel.

Q: There are always new designs for foundation wall systems on the market. What should the building official be asking for when evaluating new products?

A: The building official should ask for engineering data from a design professional and installation requirements from the manufacturer, or possibly, an ICC ES report from the ICC if one exists.

R404.1.6 Height above finished grade. Concrete and masonry foundation walls shall extend above the finished grade adjacent to the foundation at all points a minimum of 4 inches (102 mm) where masonry veneer is used and a minimum of 6 inches (152 mm) elsewhere.

Q: Other than the 4 inches for masonry veneer, are there any provisions in the code that would al-

TABLE R404.1.1(4)
12-INCH MASONRY FOUNDATION WALLS WITH REINFORCING WHERE d > 8.75 INCHES[a]

WALL HEIGHT	HEIGHT OF UNBALANCED BACKFILL[e]	MINIMUM VERTICAL REINFORCEMENT[b, c] — Soil classes and lateral soil load[d] (psf per foot below grade)		
		GW, GP, SW and SP soils 30	GM, GC, SM, SM-SC and ML soils 45	SC, ML-CL and inorganic CL soils 60
6 feet 8 inches	4 feet (or less)	#4 at 72″ o.c.	#4 at 72″ o.c.	#4 at 72″ o.c.
	5 feet	#4 at 72″ o.c.	#4 at 72″ o.c.	#4 at 72″ o.c.
	6 feet 8 inches	#4 at 72″ o.c.	#4 at 72″ o.c.	#5 at 72″ o.c.
7 feet 4 inches	4 feet (or less)	#4 at 72″ o.c.	#4 at 72″ o.c.	#4 at 72″ o.c.
	5 feet	#4 at 72″ o.c.	#4 at 72″ o.c.	#4 at 72″ o.c.
	6 feet	#4 at 72″ o.c.	#4 at 72″ o.c.	#5 at 72″ o.c.
	7 feet 4 inches	#4 at 72″ o.c.	#5 at 72″ o.c.	#6 at 72″ o.c.
8 feet	4 feet (or less)	#4 at 72″ o.c.	#4 at 72″ o.c.	#4 at 72″ o.c.
	5 feet	#4 at 72″ o.c.	#4 at 72″ o.c.	#4 at 72″ o.c.
	6 feet	#4 at 72″ o.c.	#4 at 72″ o.c.	#5 at 72″ o.c.
	7 feet	#4 at 72″ o.c.	#5 at 72″ o.c.	#6 at 72″ o.c.
	8 feet	#5 at 72″ o.c.	#6 at 72″ o.c.	#6 at 64″ o.c.
8 feet 8 inches	4 feet (or less)	#4 at 72″ o.c.	#4 at 72″ o.c.	#4 at 72″ o.c.
	5 feet	#4 at 72″ o.c.	#4 at 72″ o.c.	#4 at 72″ o.c.
	6 feet	#4 at 72″ o.c.	#4 at 72″ o.c.	#5 at 72″ o.c.
	7 feet	#4 at 72″ o.c.	#5 at 72″ o.c.	#6 at 72″ o.c.
	8 feet 8 inches	#5 at 72″ o.c.	#7 at 72″ o.c.	#6 at 48″ o.c.
9 feet 4 inches	4 feet (or less)	#4 at 72″ o.c.	#4 at 72″ o.c.	#4 at 72″ o.c.
	5 feet	#4 at 72″ o.c.	#4 at 72″ o.c.	#4 at 72″ o.c.
	6 feet	#4 at 72″ o.c.	#5 at 72″ o.c.	#5 at 72″ o.c.
	7 feet	#4 at 72″ o.c.	#5 at 72″ o.c.	#6 at 72″ o.c.
	8 feet	#5 at 72″ o.c.	#6 at 72″ o.c.	#6 at 56″ o.c.
	9 feet 4 inches	#6 at 72″ o.c.	#6 at 48″ o.c.	#6 at 40″ o.c.
10 feet	4 feet (or less)	#4 at 72″ o.c.	#4 at 72″ o.c.	#4 at 72″ o.c.
	5 feet	#4 at 72″ o.c.	#4 at 72″ o.c.	#4 at 72″ o.c.
	6 feet	#4 at 72″ o.c.	#5 at 72″ o.c.	#5 at 72″ o.c.
	7 feet	#4 at 72″ o.c.	#6 at 72″ o.c.	#6 at 72″ o.c.
	8 feet	#5 at 72″ o.c.	#6 at 72″ o.c.	#6 at 48″ o.c.
	9 feet	#6 at 72″ o.c.	#6 at 56″ o.c.	#6 at 40″ o.c.
	10 feet	#6 at 64″ o.c.	#6 at 40″ o.c.	#6 at 32″ o.c.

For SI: 1 inch = 25.4 mm, 1 foot = 304.8 mm, 1 pound per square foot per foot = 0.157 kPa/mm.

a. Mortar shall be Type M or S and masonry shall be laid in running bond.

b. Alternative reinforcing bar sizes and spacings having an equivalent cross-sectional area of reinforcement per lineal foot of wall shall be permitted provided the spacing of the reinforcement does not exceed 72 inches.

c. Vertical reinforcement shall be Grade 60 minimum. The distance from the face of the soil side of the wall to the center of vertical reinforcement shall be at least 8.75 inches.

d. Soil classes are in accordance with the Unified Soil Classification System and design lateral soil loads are for moist conditions without hydrostatic pressure. Refer to Table R405.1.

e. Unbalanced backfill height is the difference in height between the exterior finish ground level and the lower of the top of the concrete footing that supports the foundation wall or the interior finish ground levels. Where an interior concrete slab-on-grade is provided and in contact with the interior surface of the foundation wall, measurement of the unbalanced backfill height is permitted to be measured from the exterior finish ground level to the top of the interior concrete slab is permitted.

low a concrete or masonry foundation to be less than 6 inches above the adjacent grade?

A: Only if the entire wall section complies with the provisions for wood foundations in Section R404.2.

R404.3 Wood sill plates. Wood sill plates shall be a minimum of 2-inch by 4-inch (51 mm by 102 mm) nominal lumber. Sill plate anchorage shall be in accordance with Sections R403.1.6 and R602.11.

Q: Section R602.3.4, "Bottom (sole) plate," states that "studs shall have full bearing on a nominal 2 by or larger plate or sill having a width at least equal to the width of the stud." The code does not appear to address the situation where a foundation sill plate overhangs the concrete or masonry foundation wall that it rests on. Is this addressed in the code?

A: The code would imply that the bearing be carried down full width under the provisions in Section R301.1 where it addresses "complete load path." In the WFCM, Section 2.4.1.3, "Bottom Plate," states that bottom plates that are connected directly to the foundation shall have full bearing on the foundation. See Figure 4-7.

FOUNDATION SILL PLATE OVERHANGING CONCRETE FOUNDATION WALL FIGURE 4-7

SECTION R405 FOUNDATION DRAINAGE

R405.1 Concrete or masonry foundations. Drains shall be provided around all concrete or masonry foundations that retain earth and enclose habitable or usable spaces located below grade. Drainage tiles, gravel or crushed stone drains, perforated pipe or other approved systems or materials shall be installed at or below the area to be protected and shall discharge by gravity or mechanical means into an approved drainage system. Gravel or crushed stone drains shall extend at least 1 foot (305 mm) beyond the outside edge of the footing and 6 inches (152 mm) above the top of the footing and be covered with an approved filter membrane material. The top of open joints of drain tiles shall be protected with strips of building paper, and the drainage tiles or perforated pipe shall be placed on a minimum of 2 inches (51 mm) of washed gravel or crushed rock at least one sieve size larger than the tile joint opening or perforation and covered with not less than 6 inches (152 mm) of the same material.

> **Exception:** A drainage system is not required when the foundation is installed on well-drained ground or sand-gravel mixture soils according to the Unified Soil Classification System, Group I Soils, as detailed in Table R405.1.

Q: Because of the soil conditions in our area all basement foundations will need to have a drain around the foundation. Most of the builders have chosen to use 4-inch perforated drainage pipe. Could this be installed on the interior of the foundation wall just inside the exterior wall footing?

A: The primary drain system should be installed on the exterior side of the foundation wall. The main purpose of this drain system is to provide a means to divert exterior rain water away from the structure. Any interior drain system would be considered a secondary system. See Figure 4-8.

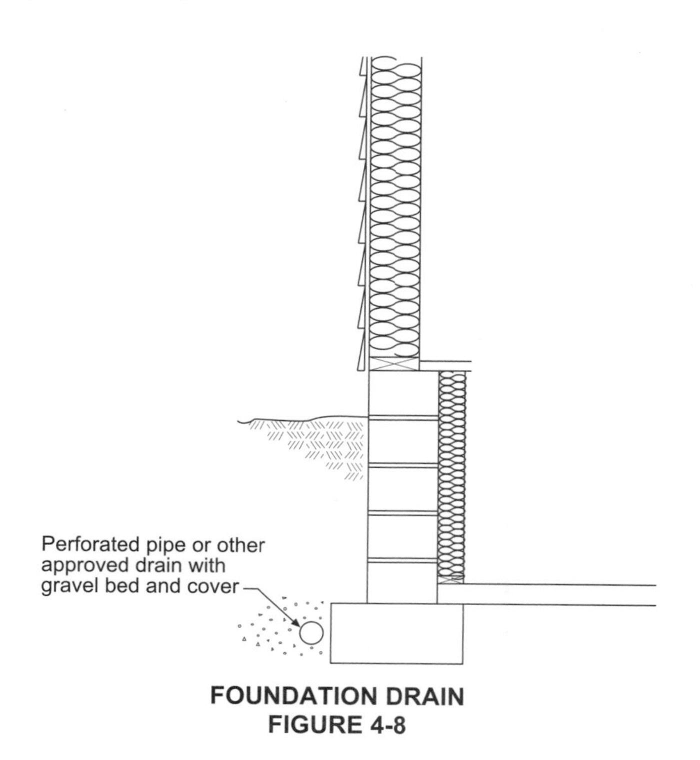

FOUNDATION DRAIN FIGURE 4-8

CHAPTER

FLOORS

5

SECTION R502
WOOD FLOOR FRAMING

R502.1 Identification. Load-bearing dimension lumber for joists, beams and girders shall be identified by a grade mark of a lumber grading or inspection agency that has been approved by an accreditation body that complies with DOC PS 20. In lieu of a grade mark, a certificate of inspection issued by a lumber grading or inspection agency meeting the requirements of this section shall be accepted.

Q: What is dimension lumber?

A: Dimension lumber refers to solid sawn lumber that is 2 inches to 4 inches thick (based on nominal dimensions) and 2 inches or more in width.

Q: What information is normally on a grade mark?

A: A grade mark should note the lumber species, grade and moisture content at the time of surfacing, grading agency, mill name and the grade's identification number.

Q: On the grade stamp it states "S-Dry." What does this refer to?

A: "S-Dry" refers to wood that has a 19-percent maximum moisture content at the time of surfacing. Also, "MC-15" refers to a 15-percent maximum moisture content and "S-GRN" refers to a moisture content over 19 percent (unseasoned wood) at the time of surfacing.

Q: What is MSR lumber?

A: "MSR" refers to machine stress-rated lumber. This is lumber that has been graded by machine stress-rated equipment instead of being visually graded. MSR lumber is typically used as components for engineered trusses but it can be used for other members with high-stress demands.

R502.1.4 Prefabricated wood I-joists. Structural capacities and design provisions for prefabricated wood I-joists shall be established and monitored in accordance with ASTM D 5055.

Q: What is ASTM D 5055?

A: ASTM D 5055 is the *Standard Specification for Establishing and Monitoring Structural Capacities of Prefabricated Wood I-Joists*. For example, the APA-Engineered Wood Association's PRI-400, *Performance Standard for APA EWS I-Joists*, includes as a referenced standard ASTM D 5055 along with other related standards. The criteria in the ASTM D 5055 standard addresses the flange material, the plywood or oriented strand board (OSB), the use of exterior-type adhesives, fasteners and other criteria.

Q: When using wood I-joists, is the sheathing required to be glued and nailed to the wood I-joists.

A: Not unless the specific application requires a glued and nailed floor system. For example, the APA PRI-400 span tables are based upon a composite floor system glued and nailed to the floor sheathing.

Q: Are squash blocks required for wood I-joists along bearing locations?

A: Only if required by the manufacturer of the I-joist. See Figure 5-1.

SQUASH BLOCKS ADJACENT TO ENGINEERED WOOD I-JOISTS
FIGURE 5-1

Q: Where does the code address the minimum end bearing for engineered floor trusses?

A: Refer to the manufacturer's installation instructions for the required end bearing for any engineered product. The end bearing is the required dimension that the floor truss needs to bear on the wall or beam below so the bearing stress is within limits.

R802.7.2 Engineered wood products. Cuts, notches and holes bored in trusses, structural composite lumber, structural glue-laminated members or I-joists are prohibited except where permitted by the manufacturer's recommendations or where the effects of such alterations are specifically considered in the design of the member by a registered design professional.

Q: Can wood I-joists be cut or beveled at the bearing locations?

A: Refer to the installation instructions of the specific manufacturer.

R502.2 Design and construction. Floors shall be designed and constructed in accordance with the provisions of this chapter, Figure R502.2 and Sections R319 and R320 or in accordance with AF&PA/NDS.

Q: We don't understand the use of the word "or" in Section R502.2. It seems to imply that the last part of the sentence is optional related to the construction of floors. Is this correct?

A: Yes. The AF&PA NDS is the engineering standard for design of wood structures referenced in the IRC. The code is a prescriptive code. It allows the use of the NDS as an alternative or when the code does not address the situation.

Q: Whose option is it to use this document?

A: It is up to the owner/designer/builder whether he or she wants to comply with the prescriptive provisions of the IRC or provides engineering according to the NDS. In addition, Section R301.1.1, "Alternate provisions," permits the use of the AF&PA WFCM as an alternative to the wood-frame construction provisions of the IRC. The NDS is an engineering standard, and the WFCM is a "how to" prescriptive manual. Both of these contain a lot of information that is in a usable format that is beneficial to the engineer, builder and building official.

Q: Can a building official require compliance with either the NDS or WFCM?

A: No, but he or she can require engineering to substantiate an installation of elements or portions that do not conform to the limitations of the IRC (see Section R301.1.3). In some cases the builder may find it easier to use the WFCM or provide engineering according to the NDS.

Q: Does the NDS have any limits on what size of structure it can apply to?

A: No. The NDS is a general engineering design standard for wood structures and is unrelated to the occupancy or use of the structure. However, the WFCM is limited to one- and two-family dwellings and has other limitations pertaining to building dimensions, mean roof height and wall height.

R502.2.1 Framing at braced wall lines. A load path for lateral forces shall be provided between floor framing and braced wall panels located above or below a floor, as specified in Section R602.10.8.

Q: What does the code require for a load path at braced wall panels?

A: Section R602.10.8 prescribes connections of braced wall lines to the framing above and below. Braced wall line sole plates must be fastened to the floor framing and top plates shall be connected to the framing above in accordance with Table R602.3(1). Sills must be fastened to the foundation or slab in accordance with Sections R403.1.6 and R602.11. Where joists are perpendicular to the braced wall lines above, the code requires blocking under and in line with the braced wall panels. Where joists are perpendicular to braced wall lines below, blocking over and in line with the braced wall panels must be provided. Where joists are parallel to braced wall lines above or below, a parallel joist must be provided at the wall and fastened in accordance with Table R602.3(1). These connections are intended to provide the required load path for the wall bracing.

R502.2.2 Decks. Where supported by attachment to an exterior wall, decks shall be positively anchored to the primary structure and designed for both vertical and lateral loads as applicable. Such attachment shall not be accomplished by the use of toenails or nails subject to withdrawal. Where positive connection to the primary building structure cannot be verified during inspection, decks shall be self- supporting. For decks with cantilevered framing members, connections to exterior walls or other framing members, shall be designed and constructed to resist uplift resulting from the full live load specified in Table R301.5 acting on the cantilevered portion of the deck.

Q: When using bolts or lag screws to attach a ledger board for a deck to the dwelling, does the code have any prescriptive information to guide the user related to size and spacing of those anchoring devices?

A: Table R301.5 establishes the live load for the deck at 40 psf. Once the total load along the ledger is established, the bolts or lag screws need to be installed based on the design capacity of the particular bolt or lag screw. See Figure 5-2.

ANCHORING OF DECK LEDGER TO DWELLING
FIGURE 5-2

R502.7 Lateral restraint at supports. Joists shall be supported laterally at the ends by full-depth solid blocking not less than 2 inches (51 mm) nominal in thickness; or by attachment to a full-depth header, band or rim joist, or to an adjoining stud or shall be otherwise provided with lateral support to prevent rotation.

> **Exception:** In Seismic Design Categories D_0, D_1 and D_2, lateral restraint shall also be provided at each intermediate support.

R502.7.1 Bridging. Joists exceeding a nominal 2 inches by 12 inches (51 mm by 305 mm) shall be supported laterally by solid blocking, diagonal bridging (wood or metal), or a continuous 1-inch-by-3-inch (25.4 mm by 76 mm) strip nailed across the bottom of joists perpendicular to joists at intervals not exceeding 8 feet (2438 mm).

Q: Do 2 x 10 floor joists require mid-span bridging or blocking?

A: No. Section R502.7.1 addresses requirements for lumber joists greater than 2 x 12. 2 x 10 and 2 x 12 floor joists do not require mid-span bridging or blocking because at least one edge is held in line for its entire length, based on the fastening of the floor sheathing on the top edge. Joists exceeding 2 x 12 must be supported laterally by solid blocking, diagonal bridging or a continuous 1 x 3 strip perpendicular to the joist bottom at intervals not exceeding 8 feet.

R502.12 Draftstopping required. When there is usable space both above and below the concealed space of a floor/ceiling assembly, draftstops shall be installed so that the area of the concealed space does not exceed 1,000 square feet (92.9 m^2). Draftstopping shall divide the concealed space into approximately equal areas. Where the assembly is enclosed by a floor membrane above and a ceiling membrane below draftstopping shall be provided in floor/ceiling assemblies under the following circumstances:

1. Ceiling is suspended under the floor framing.
2. Floor framing is constructed of truss-type open-web or perforated members.

Q: Is draftstopping required in an open web floor truss system above a basement level that exceeds 1,000 square feet if there is no finished ceiling applied to the bottom of the floor trusses?

A: No. Once the ceiling is applied to the bottom of these open web floor trusses, creating the concealed space, then draftstopping is required. See Figure 5-3.

OPEN WEB FLOOR TRUSS SYSTEM
FIGURE 5-3

R503.2 Wood structural panel sheathing.

R503.2.1 Identification and grade. Wood structural panel sheathing used for structural purposes shall conform to DOC PS 1, DOC PS 2 or, when manufactured in Canada, CSA 0437 or CSA 0325. All panels shall be identified by a grade mark of certificate of inspection issued by an approved agency.

Q: What are PS 1 and PS 2?

A: They are *Voluntary Product Standard PS 1, Construction and Industrial Plywood,* and *Voluntary Product Standard PS 2, Performance Standard for Wood-Based Structural-Use Panels.* These standards were developed jointly by the U.S. Department of Commerce and the wood structural panel industry. Plywood, OSB and composite wood fiber panels are examples of structural-use panels.

Q: Is floor sheathing required to have a minimum 1/8-inch gap between panels?

A: The code does not contain any prescriptive requirement other than to refer to the manufacturer's installation instructions. Usually, if the manufacturer requires a gap, it will be stamped on the sheathing. It will say something like "sized for spacing" or "maintain 1/8-inch gap at all edges." See Figure 5-4.

Q: Is floor sheathing required to be laid in a staggered pattern?

A: No.

SPACE BETWEEN SHEETS OF WOOD STRUCTURAL PANELS
FIGURE 5-4

Q: Is floor sheathing required to be edge blocked?

A: Yes, except for tongue-and-groove sheathing material or where underlayment is provided in accordance with Table R503.2.1.1(1).

SECTION R506
CONCRETE FLOORS (ON GROUND)

R506.1 General. Concrete slab-on-ground floors shall be a minimum 3.5 inches (89 mm) thick (for expansive soils, see Section R403.1.8). The specified compressive strength of concrete shall be as set forth in Section R402.2.

R506.2 Site preparation. The area within the foundation walls shall have all vegetation, top soil and foreign material removed.

R506.2.1 Fill. Fill material shall be free of vegetation and foreign material. The fill shall be compacted to assure uniform support of the slab, and except where approved, the fill depths shall not exceed 24 inches (610 mm) for clean sand or gravel and 8 inches (203 mm) for earth.

R506.2.2 Base. A 4-inch-thick (102 mm) base course consisting of clean graded sand, gravel, crushed stone or crushed blast-furnace slag passing a 2-inch (51 mm) sieve shall be placed on the prepared subgrade when the slab is below grade.

Exception: A base course is not required when the concrete slab is installed on well-drained or sand-gravel mixture soils classified as Group I according to the United Soil Classification System in accordance with Table R405.1.

Q: Does the code specify a minimum compressive strength of the concrete for garage slabs and driveways?

A: Yes. In Table R402.2 it addresses the minimum compressive strength of the concrete based upon specific location and weathering potential.

Q: Is steel reinforcing required by the code for the concrete floors of attached garages with independent perimeter foundation walls or for driveways?

A: No.

R506.2.3 Vapor retarder. A 6 mil (0.006 inch; 152 μm) polyethylene or approved vapor retarder with joints lapped not less than 6 inches (152 mm) shall be placed between the concrete floor slab and the base course or the prepared subgrade where no base course exists.

Exception: The vapor retarder may be omitted:

1. From garages, utility buildings and other unheated accessory structures.
2. From driveways, walks, patios and other flatwork not likely to be enclosed and heated at a later date.
3. Where approved by the building official, based on local site conditions.

Q: When the vapor retarder is installed in a basement, is it required to be sealed or taped at the joints or where it meets the exterior foundation wall?

A: No.

R506.2.4 Reinforcement support. Where provided in slabs on ground, reinforcement shall be supported to remain in place from the center to upper one third of the slab for the duration of the concrete placement.

Q: When welded wire fabric reinforcing is installed in a garage slab, are reinforcing chairs required to keep it in the middle of the slab?

A: Yes. The code now requires some method to ensure that the wire mesh is not laying against the bottom of the slab as it is in many cases.

NOTES

CHAPTER

WALL CONSTRUCTION

6

SECTION R601 GENERAL

R601.1 Application. The provisions of this chapter shall control the design and construction of all walls and partitions for all buildings.

R601.2 Requirements. Wall construction shall be capable of accommodating all loads imposed according to Section R301 and of transmitting the resulting loads to the supporting structural elements.

Q: The provisions and tables in the code seem to imply the use of uniform loads for wood-framed structures not exceeding three stories. Does the code require that walls supporting concentrated loads from large girder trusses be designed by an engineer?

A: When the structural elements exceed the limits of Section R301, those elements shall be designed in accordance with accepted engineering practice to demonstrate compliance of the nonconventional elements with the conventionally framed system. Generally, accepted engineering practice means the engineered analysis is based on well-established principles of mechanics and conforms to accepted principles, tests or standards of nationally recognized technical authority. The building official has the authority to require that it be designed and certified by an architect or engineer.

SECTION R602 WOOD WALL FRAMING

R602.1 Identification. Load-bearing dimension lumber for studs, plates and headers shall be identified by a grade mark of a lumber grading or inspection agency that has been approved by an accreditation body that complies with DOC PS 20. In lieu of a grade mark, a certification of inspection issued by a lumber grading or inspection agency meeting the requirements of this section shall be accepted.

Q: What is DOC PS 20?

A: The U.S. Department of Commerce Product Standard (PS) 20 is the *American Softwood Lumber Standard.* The ALSC is the standing committee that is responsible for maintaining PS 20. The ALSC is comprised of manufacturers, distributors, users and consumers of lumber that administers an accreditation program for the grade marking of lumber produced under the system. The ALSC in accordance with the *Procedures for the Development of Voluntary Product Standards* of the U.S. Department of Commerce and through a consensus process establishes sizes, inspection provisions, grade marking requirements and the policies and enforcement regulations for the accreditation program.

Q: Are the horizontal joints of plywood or OSB wall sheathing required to be blocked and nailed?

A: Only where specifically required by Section R602.10.7 for braced wall panel construction. See Note a of Table R602.3(3) below.

R602.3.4 Bottom (sole) plate. Studs shall have full bearing on a nominal 2-by (38 mm) or larger plate or sill having a width at least equal to the width of the studs.

Q: When a bottom plate bears on a concrete or masonry foundation wall, can the plate overhang the edge of the foundation?

A: No, full bearing should be provided. See the Q & A for Section R404.3.

R602.10 Wall bracing. All exterior walls shall be braced in accordance with this section. In addition, interior braced wall lines shall be provided in accordance with Section R602.10.1.1. For buildings in Seismic Design Categories D_0, D_1 and D_2, walls shall be constructed in accordance with the additional requirements of Sections R602.10.9, R602.10.11, and R602.11.

TABLE R602.3(3)
WOOD STRUCTURAL PANEL WALL SHEATHING

PANEL SPAN RATING	PANEL NOMINAL THICKNESS (inch)	MAXIMUM STUD SPACING (inches)	
		Siding nailed to:[a]	
		Stud	Sheathing
12/0, 16/0, 20/0, or wall —16 o.c.	$^5/_{16}$, $^3/_8$	16	16[b]
24/0, 24/16, 32/16 or wall—24 o.c.	$^3/_8$, $^7/_{16}$, $^{15}/_{32}$, $^1/_2$	24	24[c]

For SI: 1 inch = 25.4 mm.
a. Blocking of horizontal joints shall not be required.
b. Plywood sheathing $^3/_8$-inch thick or less shall be applied with long dimension across studs.
c. Three-ply plywood panels shall be applied with long dimension across studs.

R602.10.6.2 Alternate braced wall panel adjacent to a door or window opening. Alternate braced wall panels constructed in accordance with one of the following provisions are also permitted to replace each 4 feet (1219 mm) of braced wall panel as required by Section R602.10.4 for use adjacent to a window or door opening with a full-length header.

Q: Our department understands that the provisions listed in Section R602.10.6.2 for alternate braced wall panels adjacent to a door or window opening can only be used if all sheathable walls are covered with wood structural panel sheathing. Is this correct?

A: No. The provisions in this section allow the balance of the sheathable walls to be sheathed with any of the exterior wall covering materials permitted by the code. See Figure 6-1 for details of portal frame bracing.

Q: For the "portal frame" in Figure 6-2, can staples be substituted for the 8d nails at the front garage wall?

A: No. Although the code allows the substitution of staples for nails in other sheathing applications, in this particular system, it does not allow staples to be used.

NARROW WALL BRACING ADJACENT DOORS AND WINDOWS
FIGURE 6-2

SECTION R604
WOOD STRUCTURAL PANELS

R604.1 Identification and grade. Wood structural panels shall conform to DOC PS 1 or DOC PS 2. All panels shall be identified by a grade mark or certificate of inspection issued by an approved agency.

R604.2 Allowable spans. The maximum allowable spans for wood structural panel wall sheathing shall not exceed the values set forth in Table R602.3(3).

R604.3 Installation. Wood structural panel wall sheathing shall be attached to framing in accordance with Table R602.3(1). Wood structural panels marked Exposure 1 or Exterior are considered water-repellent sheathing under the code.

Q: What does the "Exposure 1" rating refer to?

A: Exposure 1 panels are made with waterproof glues with untreated wood that are designed for exposure to the weather during construction. They are not intended for permanent exposure to the weather. For permanent exposure to the weather, "exterior" grade panels must be used.

Q: Is plywood stronger than OSB of the same thickness?

A: PS 1 and PS 2 are U.S. Department of Commerce Voluntary Product Standards. PS 1 specifies how plywood must be manufactured in order to qualify for grademark identification. PS 2 deals with how a panel product must perform in a designated application rather than how it must be manufactured. Plywood and OSB may be produced to the PS 2 standard. OSB and many grades of plywood panels are performance rated based on performance-based standards that provide product performance baselines, such as load-carrying capacity for designated end uses. Although they may be manufactured differently, they are both considered wood structural panels with equivalent structural performance when used in accordance with the code.

SECTION R606 GENERAL MASONRY CONSTRUCTION

R606.1 General. Masonry construction shall be designed and constructed in accordance with the provisions of this section or in accordance with the provisions of ACI 530/ASCE 5/TMS 402.

R606.1.1 Professional registration not required. When the empirical design provisions of ACI 530/ASCE 5/TMS 402 Chapter 5 or the provisions of this section are used to design masonry, project drawings, typical details and specifications are not required to bear the seal of the architect or engineer responsible for design, unless otherwise required by the state law of the jurisdiction having authority.

R606.2.1 Minimum thickness. The minimum thickness of masonry bearing walls more than one story high shall be 8 inches (203 mm). Solid masonry walls of one-story dwellings and garages shall not be less than 6 inches (152 mm) in thickness when not greater than 9 feet (2743 mm) in height, provided that when gable construction is used, an additional 6

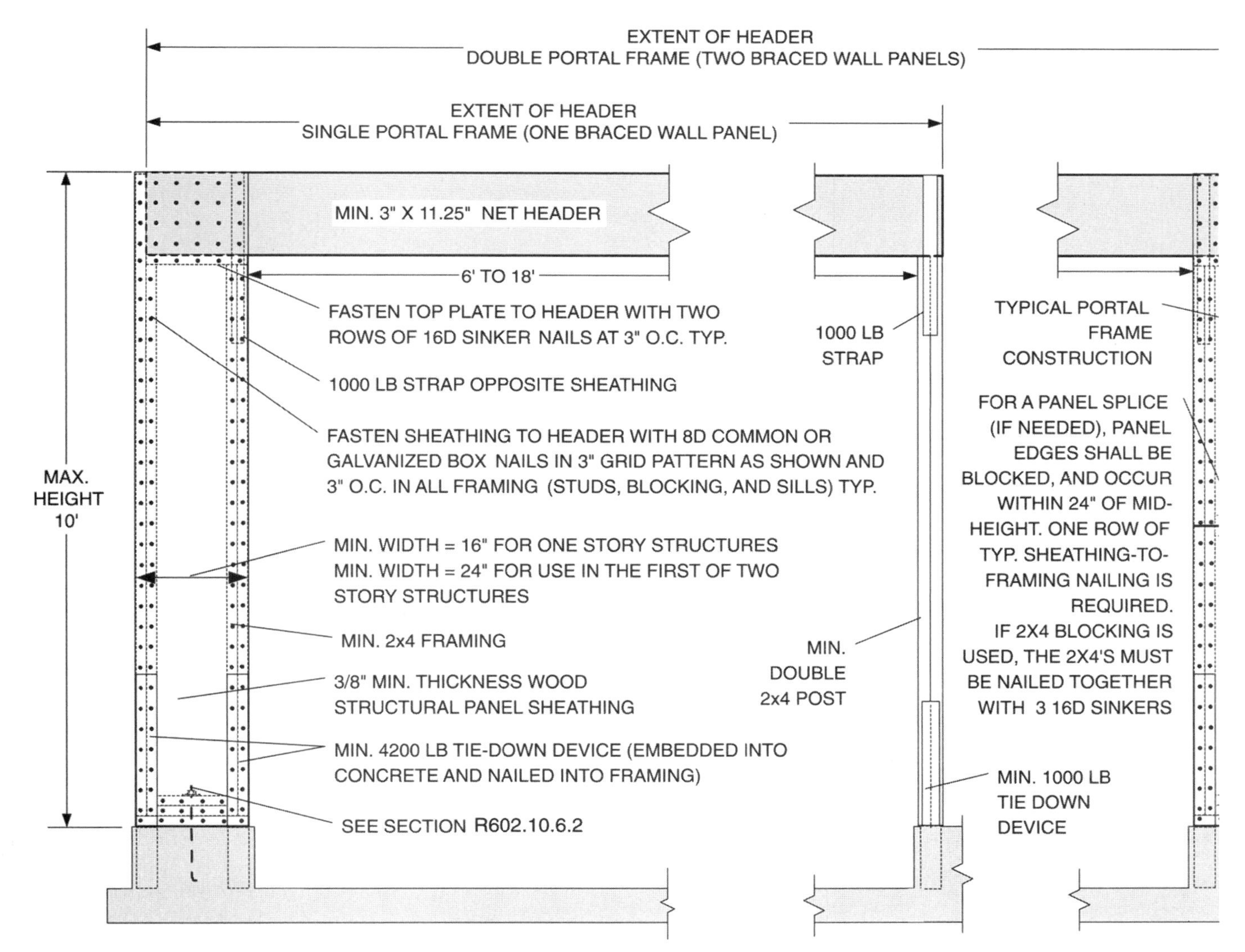

For SI: 1 inch = 25.4 mm, 1 foot = 304.8 mm, 1 pound = 0.454 kg.

ALTERNATE BRACED WALL PANEL ADJACENT TO A DOOR OR WINDOW OPENING
FIGURE 6-1 (IRC Figure R602.10.6.2)

feet (1829 mm) is permitted to the peak of the gable. Masonry walls shall be laterally supported in either the horizontal or vertical direction at intervals as required by Section R606.9.

Q: We have heard that a 4-inch masonry unit cannot be used as the top course in a garage foundation. Is this true?

A: Yes. See Section R606.2.1. Also, the code requires 1 inch of grout around all sides of the anchor bolt, and this is not possible in most 4-inch masonry units due to the lack of opening in the block itself. See Figure 6-3.

MASONRY CURB BLOCK
FIGURE 6-3

Q: What is the weight of a lightweight block?

A: Lightweight masonry units weigh 105 pcf or less.

Q: Is the top course of block in a masonry wall required to be filled solid?

A: No. See Figure 6-4.

Q: Would the top course of a 6-inch masonry wall in an attached garage need to be filled solid?

A: No.

GROUTING OF ANCHOR STRAPS IN HOLLOW MASONRY WALL
FIGURE 6-4

R606.2.3 Change in thickness. Where walls of masonry of hollow units or masonry-bonded hollow walls are decreased in thickness, a course of solid masonry shall be constructed between the wall below and the thinner wall above, or special units or construction shall be used to transmit the loads from face shells or wythes above to those below.

Q: The front walls of our basements are constructed with a 4-inch brick ledge at the top course of the masonry foundation wall. Typically, these basements will be laid up with 12-inch-wide masonry units, except for the top course being an 8-inch-wide masonry unit that provides a 4-inch brick ledge in front of it. Is the top course of the 12-inch masonry block that supports the 8-inch-wide masonry unit required to be a solid masonry unit or at least a hollow masonry unit that is filled with mortar or grout?

A: Section R606.2.3 requires the top course of the 12-inch masonry wall to be solid masonry units. If approved by the building official, a fully grouted course of hollow units would be an acceptable alternative to solid masonry units.

Q: What is a solid masonry unit?

A: In Section R202, it defines solid masonry as a masonry unit where the net cross-sectional area of each unit in any plane parallel to the bearing surface is not less than 75 percent of its gross cross-sectional area.

SECTION R607
UNIT MASONRY

R607.1 Mortar. Mortar for use in masonry construction shall comply with ASTM C 270. The type of mortar shall be in accordance with Sections R607.1.1, R607.1.2 and R607.1.3 and shall meet the proportion specifications of Table R607.1 or the property specifications of ASTM C 270.

Q: Where are the provisions for cold weather masonry construction?

A: ASTM C 270 Section 7.5, "Climatic Conditions," states, "Unless superseded by other contractual relationships or the requirements of local building codes, cold weather masonry construction relating to mortar shall comply with the *International Masonry All-Weather Council's Guide Specification for Cold Weather Masonry Construction*, Section 04200, Article 3." Under referenced documents, it notes that the document name is slightly different, now noted as *Recommended Practices and Guide Specifications for Cold Weather Masonry Construction*. In the most recent version of ASTM C 270, a change has been made to reference a new *Hot and Cold Weather Masonry Construction Guide* because the older guide is no longer available. Either of the two versions should provide appropriate guidelines for cold weather masonry construction.

R607.1.1 Foundation walls. Masonry foundation walls constructed as set forth in Tables R404.1.1(1) through R404.1.1(4) and mortar shall be Type M or S.

Q: Does the code address tooling of mortar joints?

A: Tooling is addressed in ASTM C 270.

Q: Why is Type N mortar not permitted for use in foundation walls?

A: Type N mortar has a very low compressive strength at approximately 750 psi.

R607.2 Placing mortar and masonry units.

R607.2.1 Bed and head joints. Unless otherwise required or indicated on the project drawings, head and bed joints shall be $^3/_8$ inch (10 mm) thick, except that the thickness of the bed joint of the starting course placed over foundations shall not be less than $^1/_4$ inch (7 mm) and not more than $^3/_4$ inch (19 mm).

Q: What is the height of a modular 8-inch block?

A: An 8-inch-high modular block is 7 $^5/_8$ inches. With the $^3/_8$-inch mortar joint, it works out to even 8-inch dimensions for the designer and builder.

Q: What are the compressive strengths of Type M and Type S mortar?

A: Type M mortar is approximately 2,500 psi and Type S mortar is approximately 1,800 psi.

SECTION R609
GROUTED MASONRY

R609.1 General. Grouted multiple-wythe masonry is a form of construction in which the space between the wythes is solidly filled with grout. It is not necessary for the cores of masonry units to be filled with grout. Grouted hollow unit masonry is a form of construction in which certain cells of hollow units are continuously filled with grout.

R609.1.1 Grout. Grout shall consist of cementitious material and aggregate in accordance with ASTM C 476 and the proportion specifications of Table R609.1.1. Type M or Type S mortar to which sufficient water has been added to produce pouring consistency can be used as grout.

Q: What is the minimum required compressive strength of grout?

A: Grout shall have a minimum of 2,000 psi compressive strength.

SECTION R611
INSULATING CONCRETE FORM
WALL CONSTRUCTION

R611.1 General. Insulating Concrete Form (IFC) walls shall be designed and constructed in accordance with the provisions of this section or in accordance with the provisions of ACI 318. When ACI 318 or the provisions of this section are used to design insulating concrete form walls, project drawings, typical details and specifications are not required to bear the seal of the architect or engineer responsible for design, unless otherwise required by the state law of the jurisdiction having authority.

R611.6.3 Insulation materials. Insulating concrete forms material shall meet the surface burning characteristics of

Section R314.3. A thermal barrier shall be provided on the building interior in accordance with Section R314.4 or Section R702.3.4.

Q: Does the exposed foam plastic on the interior side of the insulating concrete form wall need to be protected with ½-inch-thick gypsum board or other approved equivalent thermal barrier?

A: Yes. The references send the user back to the foam plastic provisions in Section R314 that require the thermal protection that you reference. In Section R702.3.4, it does require the foam "on the interior of habitable spaces shall be covered in accordance with Section R314.4." The text in this section is misleading when the word "habitable space" is used. The intent was to require a thermal barrier on the interior of building spaces. The foam plastic needs to meet these same requirements for thermal protection.

SECTION R613
EXTERIOR WINDOWS AND GLASS DOORS

R613.1 General. This section prescribes performance and construction requirements for exterior window systems installed in wall systems. Windows shall be installed and flashed in accordance with the manufacturer's written installation instructions. Written installation instructions shall be provided by the manufacturer for each window.

Q: In Section R703.8 the code addresses flashing of exterior window and door openings. I understand that the code now requires the manufacturer to provide installation details for a prescriptive means to flash a window. Where is this noted in the code?

A: Section R613.1 requires the window to be installed according to the written installation instructions provided by the window manufacturer. These installation instructions should be available to the building inspector on site.

R613.2 Window sills. In dwelling units, where the opening of an operable window is located more than 72 inches (1829 mm) above the finished grade or surface below, the lowest part of the clear opening of the window shall be a minimum of 24 inches (610 mm) above the finished floor of the room in which the window is located. Glazing between the floor and 24 inches (610 mm) shall be fixed or have openings through which a 4-inch-diameter (102 mm) sphere cannot pass.

Exceptions:

1. Windows whose openings will not allow a 4-inch-diameter (102 mm) sphere to pass through the opening when the opening is in its largest opened position.
2. Openings that are provided with window guards that comply with ASTM F 2006 or F 2090.

Q: An openable window is installed where the bottom of the window opening is less than 24 inches above the floor and the sill is greater than 72 inches above the exterior grade. The owner chooses to install a guard. How high above the floor must this guard extend?

A: The code does not specify a guard height. An opening that is closer than 24 inches requires protection. Construction of an on-site guard as an alternative to the window sill height requirement would have to be approved by the building official as an alternative method. It would be reasonable to permit a barrier that covers the open portion of the window, or a barrier that is 24 inches high providing an equivalent to the 24-inch sill requirement.

Q: If a second floor sleeping room emergency escape and rescue window had a sill height lower than 24 inches above the floor, and a guard was placed in front of the open portion less than 24 inches above the floor, could this window still comply with the net clear opening sizes required by Section R310.1?

A: If the openable portion of the window above the guard height meets the minimum net clear opening dimensions required in Section R310.1, the window will still satisfy the emergency escape and rescue opening requirements. Also, if a removable guard is placed in front of the window that meets the requirements in Section R310.4, the window opening will comply with Section R310.1.

WALL COVERING

CHAPTER

7

SECTION R701 GENERAL

R701.1 Application. The provisions of this chapter shall control the design and construction of the interior and exterior wall covering for all buildings.

R701.2 Installation. Products sensitive to adverse weather shall not be installed until adequate weather protection for the installation is provided. Exterior sheathing shall be dry before applying exterior cover.

Q: In our area of the country, it is not unusual to have it snow one day, have it melting in the 40°F heat the next day or to just have extended periods of heavy rain, resulting in very wet framing material. Are there special considerations related to the wet lumber, insulation products or the interior drywall?

A: Enough time should be allowed for the framing to sufficiently dry before the installation of insulation products. The manufacturers of any engineered wood products may also have special requirements for their products that should be verified. Most engineered wood products are manufactured with the assumption that the material will not exceed 19-percent moisture content during the life of the product. If that moisture level is exceeded for a period of time, it could also result in the failure of the product to perform as designed. For insulation products, especially cellulose and rigid and spray-applied foams, the moisture level inside of the stud wall cavity is critical in the correct application of the material, and steps should be taken by the installer to follow the manufacturer's guidelines. Section 702.3.5 requires that interior gypsum board not be installed where it is exposed to the weather or to water, which means the exterior weather barrier materials need to be in place prior to the installation.

R702.3.4 Insulating concrete form walls. Foam plastics for insulating concrete form walls constructed in accordance with Sections R404.4 and R611 on the interior of habitable spaces shall be covered in accordance with Section R314.4. Use of adhesives in conjunction with mechanical fasteners is permitted. Adhesives used for interior and exterior finishes shall be compatible with the insulating form materials.

Q: When an insulated concrete form wall is constructed, is the exposed foam insulation required to be protected with a ½-inch-thick gypsum thermal barrier?

A: Yes. Although Section R702.3.4 refers to "habitable spaces shall be covered," the provisions in Section R314.4 for thermal barriers are actually more restrictive; thus, the thermal barrier is required unless the product is listed to be installed without a thermal barrier.

SECTION R703 EXTERIOR COVERING

R703.1 General. Exterior walls shall provide the building with a weather-resistant exterior wall envelope. The exterior wall envelope shall include flashing as described in Section R703.8. The exterior wall envelope shall be designed and constructed in a manner that prevents the accumulation of water within the wall assembly by providing a water-resistant barrier behind the exterior veneer as required by Section R703.2. and a means of draining water that enters the assembly to the exterior. Protection against condensation in the exterior wall assembly shall be provided in accordance with Chapter 11 of this code.

Q: Section R703.1 seems to imply that the weather-resistive barrier is not the exterior siding, brick, stone veneer or exterior plaster, but the material behind it. Is this the case?

A: Almost all siding materials, including brick, stone veneer and exterior plaster are considered cladding materials that may serve to divert the initial rain and snow, but the barrier behind it needs to keep that water from entering the wall system. The weather barrier is a system made up of many components.

R703.2 Water-resistive barrier. One layer of No. 15 asphalt felt, free from holes and breaks, complying with ASTM D 226 for Type 1 felt or other approved water-resistive barrier shall be applied over studs or sheathing of all exterior walls. Such felt or material shall be applied horizontally, with the upper layer lapped over the lower layer not less than 2 inches (51 mm). Where joints occur, felt shall be lapped not less than 6 inches (152 mm). The felt or other approved material shall be continuous to the top of walls and terminated at penetrations and building appendages in a manner to meet the requirements of the exterior wall envelope as described in Section R703.1.

Exception: Omission of the water-resistive barrier is permitted in the following situations:

1. In detached accessory buildings.
2. Under exterior wall finish materials as permitted in Table R703.4.
3. Under paperbacked stucco lath when the paper backing is an approved weather-resistive sheathing paper.

Q: What is ASTM D 226?

A: ASTM D 226 is the *Specification for Asphalt-Saturated Organic Felt Used in Roofing and Waterproofing.* It contains the minimum requirements for classification, materials and manufacture, physical requirements, workmanship, finish, packaging and marking of these organic felts. See Figure 7-1.

ASPHALT-SATURATED FELT WEATHER BARRIER
FIGURE 7-1

Q: What does "Type I" refer to in the code section?

A: Type I is commonly called Type 15 asphalt felt. Type II is commonly called Type 30 asphalt felt.

Q: Is the "No. 15" a designation of weight of the material.

A: At one time, No. 15 was an industry reference to the weight of the felt per 100 square feet of coverage. In the ASTM D 226 standard, No. 15, or Type I felt, has a minimum net mass weight of 11.5 pounds per 100 feet squared.

Q: Are there different grades of Type I asphalt felt?

A: Yes. There are four grades of Type I asphalt felt based on Federal Specification UU-B-790a. The grade is based on dry tensile strength, water resistance and water vapor transmission. Grade A is the most water resistant, with a 24-hour rating, and Grade D is the least water resistant, with a 10-minute rating. However, Grade D is the only grade that meets a minimum criterion for water vapor transmission at 35 grams per square meter per 24 hours. In other words, Grade D is the only breathable paper.

Q: Can any of the four grades of this Type I asphalt felt be used behind products such as brick veneer, exterior plaster (stucco) and vinyl siding?

A: Type I, Grades A, B, C and D can be installed behind brick veneer and most siding products. Exterior plaster will specifically require the vapor permeable Grade D paper. See the comments under Section R703.6.

Q: When this Type I felt paper is installed behind brick veneer or a siding product, is it required to be sealed at the lapped joints (either vertical or horizontal)?

A: No.

Q: Can the No. 15 asphalt felt paper be installed vertically, such as in Figure 7-2?

A: No. Such felt or other approved material shall be applied horizontally, with the upper layer lapped over the lower layer not less than 2 inches.

IMPROPER VERTICAL APPLICATION OF ASPHALT FELT PAPER
FIGURE 7-2

Q: We regularly see house sheathing papers installed at the same slope as the roof of the structure, yet the installers tell us that it is approved by the manufacturer of the product. Any advice?

A: Yes, check the installation requirements of the manufacturer. Most will only permit the house sheathing paper to be installed horizontally.

Q: Is there a maximum time that a house sheathing paper can be left exposed to the weather prior to installation of the exterior finish material.

A: Although it may vary from one manufacturer to the next, many of the manufacturers require that it be covered with an approved weather-resistive covering within 120 days after application.

Q: Is the felt material required to be sealed at window and door openings?

A: There are requirements in Section R703.8 that contain provisions for flashing, and exceptions to the provisions for self-flashing windows. Rain water that enters the exterior walls of a house can lead to decay of the wood structural elements and other wall materials. Every installation needs to be evaluated as a complete system. Will a particular type of window, with a particular type of nailing flange, actually serve as a type of flashing and keep water from entering at that point? Is the nailing flange integral with the window itself, or does it slide into a track in the window? The nailing flange may need some additional weather barrier material applied over it to keep the rain from entering. The junction of where the felt paper meets the window should also be verified with the manufacturer's installation instructions. See Figure 7-3.

ASPHALT PAPER AT WINDOW
FIGURE 7-3

Q: What guidelines should a building official use when trying to evaluate materials submitted as other water-resistive barrier materials?

A: The provisions for alternate materials in Section R104.11, allow the building official to consider any material not specifically prescribed in the code, as long as that material complies with the intent of the code to be at least equivalent of that prescribed in the code. The ICC ES has published *Acceptance Criteria for Water-Resistive Barriers* (AC 38) that was established for recognition of water-resistive barriers intended to perform as secondary barriers behind exterior cladding. This standard includes paper-based barriers, felt-base barriers and polymeric-based barriers. The use of other independent test data could also be used if it was demonstrated that the product can meet ASTM D 226.

Q: Our municipality has accepted ICC ES reports for some house wrap materials. Are the requirements for the actual installation of these house wraps, or house sheathing papers, going to be different than those for the No. 15 asphalt felt papers?

A: Each product needs to be installed according to the test data, the individual ICC ES report and the manufacturer's installation instructions.

Q: Our inspection department requires that a weather-resistive barrier be installed between the front concrete entry slab and the exterior wall of the dwelling, even when there is a roof above the entry. Can we require this weather barrier here?

A: Yes. Even though the concrete entry slab (and/or steps) is installed under the cover of a roof above, it may still be susceptible to water penetration from rain water or melting snow. The concrete entry slab should not be poured directly against the exposed wall sheathing. See Figure 7-4.

WEATHER-RESISTIVE BARRIER PROTECTION AT FRONT ENTRY CONCRETE LANDING
FIGURE 7-4

Q: Does wood log siding require one layer of No. 15 felt paper?

A: Although there is not a specific siding material category listed in Table R703.4, all other wood-lapped siding products require a weather-resistive barrier, such as one layer of No. 15 felt paper that you noted, and one might assume the same for the wood log siding.

Q: Is the weather-resistive barrier required on the gable end of a structure when there is only unconditioned attic space on the inside of this wall?

A: Yes. It is required over all exterior walls. This barrier is designed to protect the structure from the exterior elements.

Q: What criteria should our department use when considering a proposal to have a foam sheathing product be considered as the exterior weather-resistive barrier?

A: Most foam products are tested for flame spread and smoke-developed ratings, not as weather-barrier materials. You may want to refer to the manufacturer for any test data to indicate how the product can be installed. The ICC ES is also looking at submittals for foam products intended for exterior weather-barrier consideration. See Figure 7-5.

FOAM WALL SHEATHING
FIGURE 7-5

R703.6 Exterior plaster. Installation of these materials shall be in compliance with ASTM C 926 and ASTM C 1063 and the provisions of this code.

R703.6.1 Lath. All lath and lath attachments shall be of corrosion-resistant materials. Expanded metal or woven wire lath shall be attached with $1^1/_2$-inch-long (38 mm), 11 gage nails having a $^7/_{16}$-inch (11.1 mm) head, or $^7/_8$-inch-long (22.2 mm), 16 gage staples, spaced at no more than 6 inches (152 mm), or as otherwise approved.

R703.6.2 Plaster. Plastering with portland cement plaster shall be not less than three coats when applied over metal lath or wire lath and shall be not less than two coats when applied over masonry, concrete, pressure-preservative treated wood or decay-resistant wood as specified in Section R319.1 or gypsum backing. If the plaster surface is completely covered by veneer or other facing material or is completely concealed, plaster application need be only two coats, provided the total thickness is as set forth in Table R702.1(1).

R703.6.2.1 Weep screeds. A minimum 0.019-inch (0.5 mm) (No. 26 galvanized sheet gage), corrosion-resistant weep screed or plastic weep screed, with a minimum vertical attachment flange of $3^1/_2$ inches (89 mm) shall be provided at or below the foundation plate line on exterior stud walls in accordance with ASTM C 926. The weep screed shall be placed a minimum of 4 inches (102 mm) above the earth or 2 inches (51 mm) above paved areas and shall be of a type that will allow trapped water to drain to the exterior of the building. The weather-resistant barrier shall lap the attachment flange. The exterior lath shall cover and terminate on the attachment flange of the weep screed.

R703.6.3 Water-resistive barriers. Water-resistive barriers shall be installed as required in Section R703.2 and, where applied over wood-based sheathing, shall include a water-resistive vapor-permeable barrier with a performance at least equivalent to two layers of Grade D paper.

> **Exception:** Where the water-resistive barrier that is applied over wood-based sheathing has a water resistance equal to or greater than that of 60 minute Grade D paper and is separated from the stucco by an intervening, substantially nonwater-absorbing layer or designed drainage space.

Q: Is exterior plaster the same as stucco?

A: Yes.

Q: What products are considered wood-based sheathing?

A: Plywood, OSB and fiberboard are considered wood-based sheathing.

Q: What is the purpose of having two layers of building paper with exterior plaster?

A: The purpose of the paper is to provide a means to keep any water behind the exterior plaster from contacting the wall sheathing material. These materials need to be kept dry to prevent them from expanding and contracting, which could affect the exterior plaster, and to keep the sheathing from rotting due to excessive moisture over an extended period of time.

Q: Where does the code refer to the standard for the Grade D paper?

A: The referenced standard for the Grade D paper is actually an old FHA/HUD specification known as HUD/FHA UU-B-790A. The IRC does not currently make a specific reference to this industry standard.

Q: Why does the code specifically require Grade D paper under exterior plaster? What is Grade D paper?

A: Type I felt is produced in four grades: A, B, C and D. Grade A paper is considered to be very water vapor resistant. When "boat" tested, it will meet the minimum 24-hour rating for water permeation through the paper. It is not vapor permeable. Grade B will meet a 16-hour rating for water permeation, but it is also not vapor permeable. Grade C will meet an 8-hour rating for water permeation through paper, but it is also not vapor permeable. Grade D will meet a 10-minute rating for water permeation through paper, and it will allow water-vapor transmission at the minimum rate of 35 grams per square meter per 24 hours. So Grade D paper is used for exterior plaster because of its ability to prevent water penetration from the outside, yet allow any trapped water vapor to pass through from the inside.

Q: Is the reference in the exception to Section R703.6.3 to "60-minute Grade D paper" a change in the code? What is 60-minute paper?

A: The reference to 60 minutes is a change in this edition of the code. A typical Grade D paper will only have a 10-minute rating for water permeation in the "boat" test, that being a horizontal position. This will increase that to a 60-minute rating for water permeation and still meet the requirements for water vapor transmission of the common Grade D paper.

Q: Could a house sheathing paper meet these Grade D requirements?

A: The ICC ES has a listing of house sheathing papers that have been tested according to ICC ES *Acceptance Criteria for Water-Resistive Barriers*, also known as AC 38. One of the standards that this criteria is based upon is the Federal Specification UU-B-790a that was addressed earlier. There are currently many house sheathing papers that will meet the Grade D standard. The user should verify each one individually.

Q: Where do I find the installation requirements for metal lath?

A: The requirements are contained ASTM C 1063, *Standard Specification for Installation of Lathing and Furring to Receive Interior and Exterior Portland Cement-Based Plaster*, referred to in Section R703.6. This specification covers the minimum requirements for lathing and furring. This section also references ASTM C 926, *Standard Specification for Application of Portland Cement-Based Plaster.*

Q: The code refers to the use of $^7/_8$-inch-long 16 gage staples for the application of the lath. What is the minimum crown width?

A: Staples shall have crowns not less than ¾ inch, and shall anchor not less than three strands of lath.

Q: The code states that nails used for the application of lath need to have at least a $^7/_{16}$-inch-diameter head. Can common nails be used for the application of lath?

A: Yes. Common nails need to be bent over so they anchor not less than three strands of lath.

Q: Are the nails used to apply the lath required to be galvanized nails?

A: Yes.

Q: In Section R703.6.1, one can find the minimum requirements for the use of nails and staples for metal lath. It requires the installation at no more than 6 inches on center. Does this provision assume the 6 inches on center is into a wall stud or other framing member, and not just into the wall sheathing?

A: Yes. The code requirement for the 6-inch spacing maximum is based on the installation into the framing member.

Q: It appears that many contractors will choose to install extra nails or staples into the wall sheathing to prevent the lath from moving in and out during the application of the brown and scratch coat. Would the code prevent this practice?

A: No. Although some in the industry have suggested that installing extra nails or staples into the wall cavity space may possibly allow moisture along the shank of the nail, or in the case of cold climate areas, may allow the movement of frost into the wall cavity where it may condense, thus adding moisture into the insulated wall cavity.

Q: When a metal plaster base, such as a metal lath is applied to a structure sheathed with wood structural panel sheathing, is the lath required to be furred out away from the surface?

A: Yes. Metal plaster bases, such as metal lath, shall be furred away from the surface at least ¼ inch. Self-furring lath will meet this requirement.

Q: What are metal plaster bases?

A: Metal plaster bases include expanded metal lath or welded or woven wire lath.

Q: Does the code require the metal lath to be installed in any particular horizontal pattern?

A: Yes. The ends of adjoining sheets of metal lath shall be staggered.

Q: Are there any requirements for the exterior lath to be installed flat, plumb or true?

A: Yes. Metal lath shall be installed so that the finished plaster surfaces are true to line with an allowable tolerance of ¼ inch in 10 feet.

Q: We have some columns that will receive stucco. Can the sheets of our diamond-mesh expanded metal lath be applied vertically?

A: No. Lath shall be applied with the long dimension at right angles to the supports. See Figure 7-6.

Q: Are the ends of the metal lath required to be lapped in any way?

A: Yes. Metal lath shall be lapped ½ inch at the sides or nest the edge ribs. Where end laps occur between the framing members, the ends of the sheets of all metal lath shall be laced or wire tied with 0.0475-inch galvanized, annealed steel wire.

Q: In a wood-framed structure, such as a new house, are control joints required for the application of stucco?

A: Yes. Control joints need to be installed in the walls so an area does not exceed 144 square feet, and the distance between control joints along the walls do not exceed 18 feet. Wall and partition height door frames are considered control joints. Also, nonload-bearing members need to be separated from load-bearing members with casing beads or other means. Wall or partition height door frames shall be considered as control joints. See Figures 7-7 and 7-8.

APPLICATION OF METAL LATH
FIGURE 7-6

Q: It appears that most manufacturers of wood structural panel sheathing recommend a $^{1}/_{8}$-inch gap between sheets. Does this recommendation hold true when the panels are used behind stucco?

A: Yes. Note a of Table 3 of ASTM C 1063 states that where plywood is used for sheathing, a minimum of $^{1}/_{8}$-inch separation shall be provided between adjoining sheets to allow for expansion.

Q: What is a weep screed?

A: An accessory used to terminate portland cement-based stucco at the bottom of exterior walls. This weep screed shall have a sloped, solid or perforated ground or screed flange to facilitate the removal of moisture from the wall cavity and a vertical attachment flange not less than 3½ inches. See Figure 7-9.

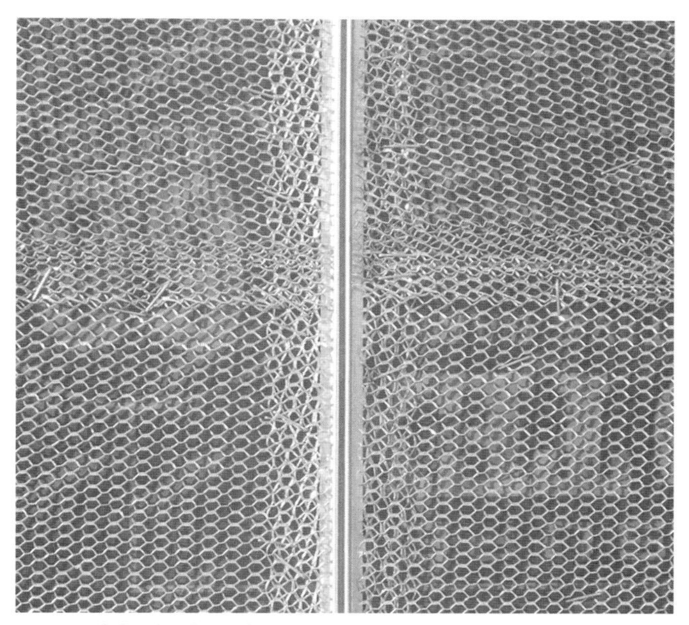

CONTROL JOINT IN EXTERIOR PLASTER
FIGURE 7-7

CONTROL JOINTS IN EXTERIOR PLASTER
FIGURE 7-8

Q: The code is specific about the location of the weep screed in relation to the grade level and paved surfaces, but where does the weep screed need to start?

A: The bottom of the weep screed needs to be installed not less than 1 inch below the point where the sill plate meets the top of the foundation wall.

WEEP SCREED
FIGURE 7-9

Q: How far does the weather-resistive barrier need to lap the attachment flange?

A: The weather-resistive barrier shall cover entirely the vertical attachment flange and terminate at the top edge of the nose or flange.

Q: Does this code section apply to variations of stucco applications, such as base coats with synthetic coatings applied as the final coat?

A: No. This section is based on ASTM C 926, and it covers the requirements for the application of full thickness portland cement plaster for exterior (stucco) work.

Q: What is a brown coat and scratch coat?

A: In a standard three-coat stucco application, a brown coat is the second coat. The scratch coat is the first coat applied to the plaster base (or lath). See Figure 7-10.

Q: Can the materials for exterior plaster be hand mixed?

A: No. All exterior plaster shall be prepared in a mechanical mixer, using sufficient water to produce a workable consistency and uniform color.

APPLICATION OF EXTERIOR PLASTER
FIGURE 7-10

Q: Can the exterior plaster mix be retempered?

A: Yes. Exterior plaster not used within 1 ½ hours from the start of initial mixing shall not be used.

Q: How soon after the application of the scratch coat can the brown coat be applied?

A: The brown (second) coat can be applied once the first coat is sufficiently rigid to support the application of the brown coat without damage to the first coat.

Q: Are there any special considerations due to extreme weather conditions?

A: Yes. Portland cement-based plaster shall not be applied to a frozen base or to a base containing frost. Plaster mixes shall not contain frozen ingredients. Plaster coats shall be protected from freezing for a period of not less than 24 hours after set has occurred. It shall also be protected from uneven and excessive evaporation during dry weather and from strong blasts of dry air.

Q: Can temporary heaters be used around stucco during application?

A: Yes. Heaters shall be located to prevent a concentration of heat on uncured plaster. Heaters shall be vented to the outside to prevent toxic fumes and other products of combustion from adhering to or penetrating plaster bases and plaster. Adequate ventilation shall be maintained in all areas, particularly in interior areas with little or no natural air movement.

Q: Is exterior plaster waterproof?

A: No. Exterior plaster (stucco) is resistant to rain penetration after it is properly cured, but is not considered waterproof.

Q: Is flashing required for exterior plaster?

A: Yes. Flashing is required at openings, perimeters and terminations to prevent water from getting behind the exterior plaster. Flashing shall be corrosion-resistant material.

Q: Can aluminum flashing be used with exterior plaster?

A: No. "Aluminum flashing shall not be used" for flashing as noted in Section A2 of ASTM C 926, *The Application of Portland Cement-Based Plaster.* The portland cement-based plaster has a very high alkaline content with a pH of 12 to 13. The aluminum reacts with alkali hydroxides in the portland cement, which will corrode aluminum.

R703.7 Stone and masonry veneer, general. Stone and masonry veneer shall be installed in accordance with this chapter, Table R703.4 and Figure R703.7. These veneers installed over a backing of wood or cold-formed steel shall be limited to the first story above-grade and shall not exceed 5 inches (127 mm) in thickness.

Q: Is stone and masonry veneer the same as adhered stone and masonry veneer?

A: No. Stone and masonry veneer are either natural or manufactured products that typically bear on a brick ledge, and are held to the structure with a series of mechanical ties. They are not bonded to the backing material. Adhered veneer products are typically applied over a portland cement base, metal lath and proper weather barrier. See Figure 7-11.

ANCHORED BRICK VENEER
FIGURE 7-11

Q: Are the steel lintels used to support stone or masonry required to be primed or painted?

A: Yes. Although not specifically noted in the IRC, in Sections 2103 and 2203 of the IBC, the steel would be required to be corrosion resistant, or be protected against corrosion with an approved coat of paint, enamel or other approved protection. See Figure 7-12.

R703.7.4 Anchorage. Masonry veneer shall be anchored to the supporting wall with corrosion-resistant metal ties. Where veneer is anchored to wood backings by corrugated sheet metal ties, the distance separating the veneer from the sheathing material shall be a maximum of a nominal 1 inch (25 mm). Where the veneer is anchored to wood backings using metal strand wire ties, the distance separating the veneer from the sheathing material shall be a maximum of $4^1/_2$ inches (114 mm). Where the veneer is anchored to cold-formed steel backings, adjustable metal strand wire ties shall be used. Where veneer is anchored to cold-formed steel backings, the distance separating the veneer from the sheathing material shall be a maximum of $4^1/_2$ inches (114 mm).

Q: When anchoring brick veneer, what does the code require when sheet metal ties are installed?

A: In Section R703.7.4.1, sheet metal ties "shall not be less than No. 22 US gage by $^7/_8$ inch corrugated. Each tie shall be spaced not more than 24 inches on center horizontally and vertically and shall support not more than 2.67 square feet of wall area."

WEATHER PROTECTION FOR STEEL LINTELS
FIGURE 7-12

R703.8 Flashing. Approved corrosion-resistant flashing shall be applied shingle-fashion in such a manner to prevent entry of water into the wall cavity or penetration of water to the building structural framing components. The flashing shall extend to the surface of the exterior wall finish. Approved corrosion-resistant flashings shall be installed at all of the following locations:

1. Exterior window and door openings. Flashing at exterior window and door openings shall extend to the surface of the exterior wall finish or to the water-resistive barrier for subsequent drainage.
2. At the intersection of chimneys or other masonry construction with frame or stucco walls, with projecting lips on both sides under stucco copings.
3. Under and at the ends of masonry, wood or metal copings and sills.
4. Continuously above all projecting wood trim.
5. Where exterior porches, decks or stairs attach to a wall or floor assembly of wood-frame construction.
6. At wall and roof intersections.
7. At built-in gutters.

Q: Section R703.8, Item 1, requires that windows and doors be flashed with a corrosion-resistant flashing. Do windows specifically require a separate flashing even though they appear to be designed as self-flashing?

A: In Section R613.1, the code requires that "windows shall be installed and flashed in accordance with the manufacturer's written installation instructions." Prior to the 2006 edition of the IRC, this language did not appear in the code, and it was left up to the field inspector to try to determine what was needed for a particular brand of windows to be considered properly flashed. Many windows were manufactured with slide-in and snap-in nailing flanges, integral nailing flanges, no nailing flanges, open corners and other variables that made it nearly impossible for the user to determine if a separate piece of flashing needed to be installed. This new language will require the manufacturer to determine what is needed for its particular windows to keep the water to the exterior.

Q: Does the code still recognize the use of caulking or J-channels above wood trim?

A: The code specifically requires flashing in this situation because the top of the wood trim is normally flat. If the caulking was not maintained, it would allow water to penetrate inward toward the structure. A J-channel would not normally project over the wood trim, and water can flow under the J-channel into the structure.

Q: What type of flashing is appropriate for flashing a deck rim to the dwelling?

A: The code would require that an approved corrosion-resistant flashing, such as hot-dipped galvanized, silicon bronze or copper, be installed in such a way as to prevent the entry of water to the building structural frame. This may include a combination of flashing materials that would be appropriate for the particular situation. When using pressure-preservative-treated wood, aluminum shall not be in contact with the wood based on the ICC ES reports. See Figure 7-13.

Q: Does the code address flashing other penetrations in the exterior envelope?

A: Yes. Section R703.1 requires that the exterior walls shall provide the building with a weather-resistant exterior envelope.

FLASHING ABOVE DECK LEDGER
FIGURE 7-13

Q: We regularly see one-story attached wood-framed structures where the roof ties into the adjacent wall of a two-story structure. We understand that flashing of the roof will be by the step-flashing method. Is some sort of kick-out flashing required at the bottom where this flashing terminates into the siding, stone or stucco finish material? We have included a photograph of an example for your consideration. See Figure 7-14.

A: Although the code does not specifically require a "kick-out" flashing as you noted in your question, it does require that the weather-resistive barrier divert the water away from the wall behind it.

R703.11 Vinyl siding. Vinyl siding shall be certified and labeled as conforming to the requirements of ASTM D 3679 by an approved quality control agency.

R703.11.1 Installation. Vinyl siding, soffit and accessories shall be installed in accordance with the manufacturer's installation instructions.

Q: What is ASTM D 3679?

A: ASTM D 3679 is the *Standard Specification for Rigid Poly Vinyl Chloride (PVC) Siding.* It specifies the requirements and test methods for extruded single-wall siding manufactured from PVC compound. Siding produced to this specification shall be installed in accordance with ASTM D 4756. This standard also requires the installation of the siding product in accordance with the manufacturer's installation instructions.

KICK-OUT FLASHING
FIGURE 7-14

Q: How does the user know if a siding product meets these standards?

A: The product label needs to identify the compliance with the standard. As a service to the industry, the Vinyl Siding Institute (VSI) sponsors a program that allows manufacturers to certify, with independent, third-party verification, that their siding meets or exceeds this specification, and this information is then posted on their websites. The product will also contain a VSI-certified label.

Q: Table R703.4 contains some provisions for nails and staples used in the application of vinyl siding, but it appears to be limited to that information. Where could I obtain more specific information regarding that fastening of the vinyl siding products?

A: Most manufacturers will provide installation instructions for their products, or they may refer to the VSI *Vinyl Siding Installation Manual.* Generally, aluminum, galvanized steel or other corrosion-resistant nails, staples or screws are utilized for the installation of vinyl siding. Aluminum trim pieces should be installed using aluminum or stainless steel fasteners. Refer to each manufacturer for specific information related to its product.

Q: Should the manufacturer's installation instructions address how to install flashing, whether rigid or flexible, around a window?

A: Yes, it is now required by Section R613.1. It is also addressed in the VSI *Vinyl Siding Installation Manual.* Generally, the flashing is installed on the underside of the window first. The flashing is then installed on the sides of the window, overlapping the bottom flashing. Third, the flashing is installed at the top of the window. The flashing should extend past the nail flanges, and be long enough to direct water over the nail flange of the last course of siding.

Q: Is there a standard means to install vents in exterior walls that are to be covered with vinyl siding?

A: Each manufacturer may recommend its own method of treating the vents. Most will provide J-blocks or other projecting elements that provide a means to install a round vent opening over a flat surface, yet allowing for the installation of a flashing system around the vent to divert the water away. See Figure 7-15.

VENT PENETRATIONS
FIGURE 7-15

Q: On a recent inspection, a contractor inadvertently omitted a required exhaust vent in the vinyl siding. The contractor cut a hole through the vinyl siding and installed the vent. Now he would like us to approve the installation; see Figure 7-16. Would this meet code?

A: No. This installation will not prevent water from penetrating behind the siding product. The siding manufacturer should be contacted for recommendations.

IMPROPER VENT INSTALLATION
FIGURE 7-16

Q: Sometimes we are presented with situations where we are not sure of what to say regarding a siding installation. Figure 7-17 shows two sections of PVC pipe penetrating through the siding where we can actually see the wall sheathing. Is this a code violation or noncompliance with the manufacturer's installation instructions?

A: If you can see the wall sheathing, it does not comply with the general provisions in Section R703.1 or the manufacturer's installation instructions.

IMPROPERLY INSTALLED PVC PENETRATIONS
FIGURE 7-17

Q: Where are the code requirements for adhered veneer?

A: The requirements are referred to in Note z of Table R703.4, which states, "Adhered masonry veneer shall comply with the requirements in Sections 6.1 and 6.3 of ACI 530/ASCE 5/TMS 402."

Q: Is that a single referenced document or three separate references?

A: The American Concrete Institute (ACI), the American Society of Civil Engineers (ASCE) and The Masonry Society (TMS) created a standard titled *Building Code Requirements for Masonry Structures,* also known as ACI 530/ASCE 5/TMS 402. Section 6.3 of this document deals specifically with prescriptive requirements for adhered veneer. They are based on adhered veneer where the unit size does not exceed $2^5/_8$ inch thickness, 36 inches in any face dimension and a total face area per unit not to exceed 5 square feet.

Q: Are there any prescriptive weight limitations in the standard?

A: Yes. The weight of the veneer cannot exceed 15 psf.

Q: Does the code consider adhered masonry veneer to be the same as brick veneer?

A: No. Adhered masonry veneer is normally considered by the industry to be a nonload-bearing natural or manufactured stone, masonry or terra cotta product adhered to a metal lath with portland cement plaster.

Q: Does the adhered veneer require a felt paper behind the metal lath?

A: Section 6.3.2.3, "Backing," refers to ACI 524R, *Guide to Portland Cement Plastering for Metal Lath, Accessories, and Their Installations.* In ACI 524R, Section 5.1.1, "Expanded Metal Lath," it refers back to ASTM C 1063, the same reference noted in the code for exterior plaster (stucco). In Section 5.2, "Weather Barrier Backing," it refers to the use of two layers of Grade D paper; once again, the same as for exterior plaster.

Q: An adhered veneer is applied half-way up a wall, with the upper portion of the wall covered with a wood-siding product. Does the adhered veneer require a weep screed? See Figure 7-18.

A: Yes. The reference standard takes the user right back to the same standard as required for exterior cement plaster on lath, that being ASTM C 1063.

BASE COAT FOR ADHERED STONE VENEER
FIGURE 7-18

Q: In a situation where adhered veneer was installed half-way up a wall, and a siding product is installed above the adhered veneer, would flashing be required above the adhered veneer where it meets the siding product?

A: Yes.

NOTES

CHAPTER

8

ROOF-CEILING CONSTRUCTION

SECTION R801 GENERAL

R801.1 Application. The provisions of this chapter shall control the design and construction of the roof-ceiling system for all building

SECTION R802 WOOD ROOF FRAMING

R802.1 Identification. Load-bearing dimension lumber for rafters, trusses and ceiling joists shall be identified by a grade mark of a lumber grading or inspection agency that has been approved by an accreditation body that complies with DOC PS 20. In lieu of a grade mark, a certificate of inspection issued by a lumber grading or inspection agency meeting the requirements of this section shall be accepted.

Q: Is the solid sawn lumber used as a component in a roof truss usually machine stress-rated lumber?

A: No. The majority of engineered roof trusses are fabricated with visually graded lumber. MSR lumber may be used for truss chords that have high combined stresses. The species and grade of the truss chords and webs should be indicated on the truss drawings.

R802.1.3 Fire-retardant-treated wood. Fire-retardant-treated wood (FRTW) is any wood product which, when impregnated with chemicals by a pressure process or other means during manufacture, shall have, when tested in accordance with ASTM E 84, a listed flame spread index of 25 or less and shows no evidence of significant progressive combustion when the test is continued for an additional 20-minute period. In addition, the flame front shall not progress more than 10.5 feet (3200 mm) beyond the center line of the burners at any time during the test.

R802.1.3.2 Strength adjustments. Design values for untreated lumber and wood structural panels as specified in Section R802.1 shall be adjusted for fire-retardant- treated wood. Adjustments to design values shall be based upon an approved method of investigation which takes into consideration the effects of the anticipated temperature and humidity to which the fire-retardant-treated wood will be subjected, the type of treatment and redrying procedures.

R802.1.3.2.1 Wood structural panels. The effect of treatment and the method of redrying after treatment, and exposure to high temperatures and high humidities on the flexure properties of fire-retardant-treated softwood plywood shall be determined in accordance with ASTM D 5516. The test data developed by ASTM D 5516 shall be used to develop adjustment factors, maximum loads and spans, or both for untreated plywood design values in accordance with ASTM D 6305. Each manufacturer shall publish the allowable maximum loads and spans for service as floor and roof sheathing for their treatment.

Q: Some of the builders in our area are using fire-retardant-treated wood structural panel roof sheathing for a distance of 4 feet on each side of the common wall between townhouse units. Usually, the roof trusses are installed at 24 inches on center. Are the span ratings of wood structural panels used as roof sheathing reduced because of the treatment with the fire-retardant chemicals?

A: Generally, yes. Many of the manufacturers of the fire-retardant chemicals for the wood structural panels are currently revising their span tables based upon independent third-party testing. When updated ICC ES reports are issued for these chemical manufacturers and their treaters, the updated information will most likely be reflected in their reports. Others have already updated their websites with this information. For designers, builders and building inspectors in areas subject to heavy snow loads, they should note the snow load tables from each individual manufacturer. Some of the span tables currently available only reflect a maximum snow load of 30 psf.

R802.6 Bearing. The ends of each rafter or ceiling joist shall have not less than $1^1/_2$ inches (38 mm) of bearing on wood or metal and not less than 3 inches (76 mm) on masonry or concrete.

Q: The code addresses the end bearing of solid sawn lumber. Where does the code address the minimum required end bearing for engineered products?

A: Each truss drawing shows a support reaction (end load) and a minimum required end-bearing length. The required end bearing, usually stated in inches and fractions of an inch, is the minimum length of bearing required at the top plate based on the allowable compressive stress perpendicular to the grain of the truss bottom chord. See Figure 8-1.

TRUSS END BEARING
FIGURE 8-1

R802.10 Wood trusses.

R802.10.1 Truss design drawings. Truss design drawings, prepared in conformance to Section R802.10.1, shall be provided to the building official and approved prior to installation. Truss design drawings shall include, at a minimum, the information specified below. Truss design drawing shall be provided with the shipment of trusses delivered to the jobsite.

Q: Section R802.10.1 notes that the truss drawings shall be provided to the building official and approved prior to installation. In our city, we always require the full set of truss drawings with the submittal documents when a house permit is applied for. Does the code allow us to require these drawings?

A: The truss design drawings are certainly needed prior to the actual installation of the trusses and they are needed by the plans examiner during the review process. These drawings are critical in determining point loads that may not be evident on the original house plans. Depending upon how the truss designer lays out the roof system, it can actually change the location of the significant point loads. In the case of larger homes with longer span girder trusses, it is not unusual to have point loads in the tens of thousands of pounds. Not only does the wall framing supporting these bearing points need to be evaluated for proper support, but the foundation wall, footings and soil need to be considered. If these issues are not addressed until the framing inspection, it may be too late to provide adequate support and a continuous load path.

Q: Truss design drawings indicate the end loads for each truss and girder. Why don't the truss manufacturers indicate what required support is needed under these larger loads?

A: Each truss drawing shows a support reaction (end load) and a minimum required end-bearing length for that particular truss. The required end bearing, usually stated in inches and fractions of an inch, is the minimum bearing length that the truss requires to rest across the top plate. This is a function of the ability of the bottom chord member of the truss to support the load without crushing the fibers. It is not related to how many studs are required below the top plate to carry the load. The truss designer knows what this load is, and it is reflected on his or her truss drawing, but because the designer does not know the specifics of how the wall is constructed (such as what grade of top plate and studs), the bearing length is based on the allowable compressive stress perpendicular to the grain of the truss bottom chord. This information must be used by the person that designed the building so he or she can determine the proper bearing support needed. See Figure 8-2.

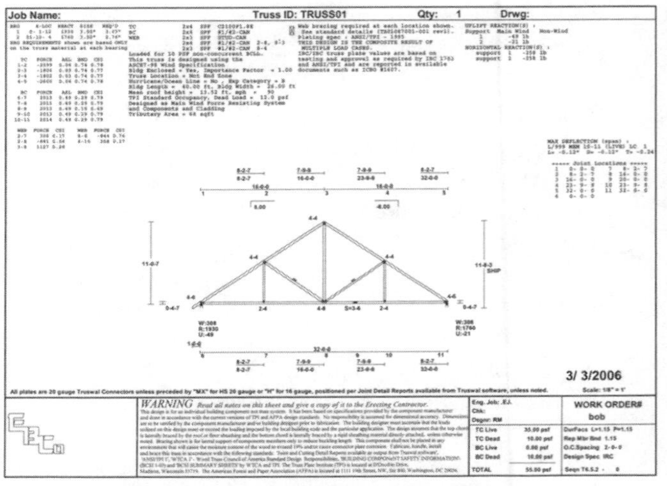

ENGINEERED TRUSS DRAWING
FIGURE 8-2

R802.10.2.1 Applicability limits. The provisions of this section shall control the design of truss roof framing when snow controls for buildings not greater than 60 feet (18 288 mm) in length perpendicular to the joist, rafter or truss span, not greater than 36 feet (10 973 mm) in width parallel to the joist span or truss, not greater than two stories in height with each story not greater than 10 feet (3048 mm) high, and roof slopes not smaller than 3:12 (25-percent slope) or greater than 12:12 (100-percent slope). Truss roof framing constructed in accordance with the provisions of this section shall be limited to sites subjected to a maximum design wind speed of 110 miles per hour (49 m/s), Exposure A, B or C, and a maximum ground snow load of 70 psf (3352 Pa). Roof snow load is to be computed as: 0.7 p_g.

Q: If a single-family dwelling falls outside of the dimension, roof slope and/or structural design criteria noted in Section R802.10.2.1 for applicability limits, does the structure then need to be designed by a registered design professional?

A: No. This only applies to wood trusses in Section R802.10, not the entire structure. The truss roof design methods are limited to the stated design criteria. Designs for truss roof framing exceeding any of the criteria have to utilize other methods such as compliance with AF&PA WFCM or an engineered design in accordance with the IBC.

R802.10.3 Bracing. Trusses shall be braced to prevent rotation and provide lateral stability in accordance with the requirements specified in the construction documents for the building and on the individual truss design drawings. In the absence of specific bracing requirements, trusses shall be braced in accordance with the Building Component Safety Information (BCSI 1-03) Guide to Good Practice for Handling, Installing & Bracing of Metal Plate Connected Wood Trusses.

Q: Will the manufacturer of the roof trusses specify the size and nailing of the required bracing?

A: Yes. It should be included on the truss drawings. See Figure 8-3.

Q: Does the code require temporary bracing for roof trusses?

A: The code requires the truss designer to address the location of permanent truss member bracing locations. Temporary bracing is in the referenced document BCSI 1-03, *Guide to Good Practice For Handling, Installing & Bracing of Metal Plate Connected Wood Trusses*, jointly produced by TPI and the Wood Truss Council of America (WTCA). The 2006 edition of BCSI replaces BCSI 1-03 and consists of 10 chapters that have been updated to include the most current information regarding the handling, installing, restraining and bracing of metal plate connected wood trusses. This edition includes information and guidance pertaining to: hoisting and placement of truss bundles; long span truss installation; design and installation of permanent bracing of individual truss members; and toe-nailed connections for attaching trusses at bearing locations.

TRUSS BRACING
FIGURE 8-3

R802.10.5 Truss to wall connection. Trusses shall be connected to wall plates by the use of approved connectors having a resistance to uplift of not less than 175 pounds (779 N) and shall be installed in accordance with the manufacturer's specifications. For roof assemblies subject to wind uplift pressures of 20 pounds per square foot (960 Pa) or greater, as established in Table R301.2(2), adjusted for height and exposure per Table R301.2(3), see section R802.11.

Q: When the code uses the word "approved connector," does it imply the use of some specific type of hardware, such as a hurricane clip?

A: The building official has the authority to approve any connector that is appropriate for the application, has adequate capacity to resist the loading requirements and meets the manufacturer's installation requirements.

Q: Could the connection between the roof truss and the top plate be accomplished by the use of nails only?

A: Depending upon the design loads, it may be possible. If you start adding up the amount of nails needed to resist the uplift, based perhaps on between 25 to 35 pounds per nail, it may get to the point where the truss is splitting due to all of the toe nails. For smaller framing members or mono trusses, it may be more practical to use nails only.

R802.10.4 Alterations to trusses. Truss members shall not be cut, notched, drilled, spliced or otherwise altered in any way without the approval of a registered design professional. Alterations resulting in the addition of load (e.g., HVAC equipment, water heater) that exceeds the design load for the truss shall not be permitted without verification that the truss is capable of supporting such additional loading.

Q: A series of roof trusses was damaged in the delivery to the job site. The contractor pieced them back together, and later called for a framing inspection. During the framing inspection the building inspector stopped the insulation contractor from installing any attic insulation or ceiling vapor barrier until a licensed structural engineer could evaluate the trusses, provide a design for the correction and have the builder call for a reinspection after completion of the corrections. The builder has suggested that we should not have to hold up the progress of the job for the truss corrections. Are we asking for too much from the builder?

A: You are probably asking for the same thing that most inspectors would ask for. The framing needs to be evaluated by the engineer and it needs to be corrected and reinspected. It also needs to remain accessible for the inspection. See Figure 8-4.

Q: We understand that the code allows substitutions of nails and staples for most wood structural panel installations. In a situation where a truss correction requires a specific size of a wood gusset, and the engineer requires it to be nailed in a particular pattern, could staples be substituted for the nails?

A: The structural engineer who designed the correction will most likely not allow substitutions of any of the materials in the correction unless specifically noted. Always verify with the engineer of record.

TRUSS ALTERATIONS
FIGURE 8-4

SECTION R803
ROOF SHEATHING

R803.1 Lumber sheathing. Allowable spans for lumber used as roof sheathing shall conform to Table R803.1. Spaced lumber sheathing for wood shingle and shake roofing shall conform to the requirements of Sections R905.7 and R905.8. Spaced lumber sheathing is not allowed in Seismic Design Category D_2.

Q: Can solid sawn lumber boards be used for sheathing a new house and garage?

A: Yes. See Table R803.1.

TABLE R803.1
MINIMUM THICKNESS OF LUMBER ROOF SHEATHING

RAFTER OR BEAM SPACING (inches)	MINIMUM NET THICKNESS (inches)
24	$^5/_8$
48[a]	$1^1/_2$ T & G
60[b]	
72[c]	

For SI: 1 inch = 25.4 mm.
a. Minimum 270 F_b, 340,000 E.
b. Minimum 420 F_b, 660,000 E.
c. Minimum 600 F_b, 1,150,000 E.

Q: Does the code address a minimum or maximum spacing of solid sawn roof boards?

A: The code does not address a minimum or maximum spacing. Industry recommends at least

$^1/_8$-inch between boards. The maximum spacing may be dictated by the allowable load capacity of the lumber itself and the ability of the roofer to put every required nail into the deck boards. See Figure 8-5.

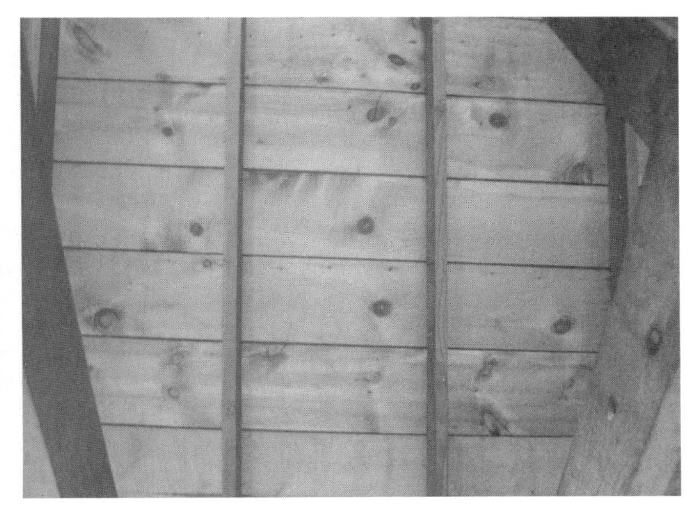

LUMBER ROOF SHEATHING
FIGURE 8-5

Q: When reroofing an older home that has board sheathing, does the code require that a layer of wood structural panel sheathing be installed over the boards prior to installing the roofing material.

A: No. Refer to the installation requirements of the shingle manufacturer. Many will allow the use of solid board sheathing with some limits on board width.

R803.2 Wood structural panel sheathing.

R803.2.1 Identification and grade. Wood structural panels shall conform to DOC PS 1, DOC PS 2 or, when manufactured in Canada, CSA 0437, and shall be identified by a grade mark or certificate of inspection issued by an approved agency. Wood structural panels shall comply with the grades specified in Table R503.2.1.1(1).

Q: We have noticed the notation "This Side Down" on wood structural panels used for roof sheathing. Is this a code requirement?

A: It is not a specific code requirement to install the sheathing with a particular side up or down. However, the manufacturer's recommendations should be followed to ensure proper performance. Some roof sheathing, such as C-D Exposure 1 plywood (commonly called "CDX" in the trade), has different veneer grades on the face and back of the panel. If the sheathing is a tongue-and-groove sheathing, it may be necessary to have all of the panels with the same orientation, such as all sheets facing down, to properly align the edge members. Also, for any type of roof sheathing, if the panels were installed "This Side Down," then the building inspector can read the grade stamp even if the underlayment or shingles are installed.

R803.2.2 Allowable spans. The maximum allowable spans for wood structural panel roof sheathing shall not exceed the values set forth in Table R503.2.1.1(1), or APA E30.

Q: When are H clips required for roof sheathing panels?

A: Table R503.2.1.1(1) contains the allowable spans for wood structural panels used for roof sheathing and floor sheathing. The span ratings are based on panel thickness, live load, total load and whether the panel has edge support or no edge support. Note d of that table describes edge support as "lumber blocking, panel edge clips (one midway between each support, except two equally spaced between supports when the span is 48 inches), tongue-and-groove panel edges, or other approved type of edge support." For some span ratings, the H clips can increase the span rating by 4 to 12 inches.

Q: When we look at the span ratings noted on the sheathing, can we assume that we also need to look at this table to verify the allowable span?

A: Yes. It is very possible that the snow load requirements in the table are more restrictive than the rating on the sheathing itself.

Q: Are the APA-rated sheathing span ratings based on a snow load of greater than 30 psf?

A: Yes. For APA-rated sheathing, live load at maximum roof spans is 30 psf for most panels, except those marked 48/24 and greater, which are 35 psf at maximum span, and 26/16 panels which are 40 psf at maximum span. These are based on a dead load of 10 pounds per square foot. It also assumes the panels are installed perpendicular to the roof framing members and across three or more supports.

Q: If we are designing with live loads (snow loads) that exceed the span ratings, are there any other options, other than having it engineered?

A: The APA Technical Note, *Load-Span Tables for APA Structural-Use Panels,* in Table 1 has a span table that covers some very large loads governed by deflection, bending and shear.

Q: Where does the code address the limitations for gable end (rake) overhangs?

A: The code does not address the situation. It is a design issue. The manufacturer of the roof sheathing requires the sheathing to be properly supported. Often, the roof sheathing is improperly installed when the carpenter cantilevers the roof sheathing about 12 inches and then fastens a fly rafter to the bottom outside edge with no apparent support. Although not specifically addressed in the code, the AF&PA WFCM (which is an optional manual based on engineering that a builder may choose to use), addresses the issue. It notes that rake overhangs shall not exceed the lesser of one-half of the purlin length or 2 feet and rake overhangs using lookout blocks shall not exceed 1 foot. See Figures 8-6 and 8-7.

GABLE-END SUPPORT
FIGURE 8-6

Q: Should the design of the particular rake detail be noted on the house plan that is included with the submittal documents?

A: Yes. If it was not framed correctly, it would most likely not be looked at by the building inspector until the framing inspection, assuming that the soffit has not already been installed also.

GABLE-END FRAMING
FIGURE 8-7

Q: We have a home under construction that has some long rafter spans that exceed the limitations of the tables in the code. These rafters have been spliced. A licensed structural engineer has stamped the plan due to the unique design. Does the IRC allow spliced rafters?

A: The tables in the code are for continuous rafters based on clear spans only. The IRC does allow engineered elements within an otherwise conventional light-frame construction. The code would require that such elements be designed in accordance with accepted engineering practice. The engineer must submit design calculations to verify that the spliced rafters will carry all applicable loads and provide details of the splice.

R803.2.3 Installation. Wood structural panel used as roof sheathing shall be installed with joints staggered or not staggered in accordance with Table R602.3(1), or APA E30 for wood roof framing or with Table R804.3 for steel roof framing.

Q: A builder is installing fire-retardant roof sheathing at the fire-resistant common walls between dwelling units in a townhouse. Is the roof sheathing required to be laid in a staggered pattern?

A: No. It is not required by the code or by the APA-The Engineered Wood Association. The APA does recommend a staggered pattern. See Figure 8-8.

PLACEMENT OF ROOF SHEATHING
FIGURE 8-8

SECTION R806
ROOF VENTILATION

R806.1 Ventilation required. Enclosed attics and enclosed rafter spaces formed where ceilings are applied directly to the underside of roof rafters shall have cross ventilation for each separate space by ventilating openings protected against the entrance of rain or snow. Ventilating openings shall be provided with corrosion-resistant wire mesh, with $^1/_8$ inch (3.2 mm) minimum to $^1/_4$ inch (6 mm) maximum openings.

R806.2 Minimum area. The total net free ventilating area shall not be less than $^1/_{150}$ of the area of the space ventilated except that reduction of the total area to $^1/_{300}$ is permitted, provided that at least 50 percent and not more than 80 percent of the required ventilating area is provided by ventilators located in the upper portion of the space to be ventilated at least 3 feet (914 mm) above the eave or cornice vents with the balance of the required ventilation provided by eave or cornice vents. As an alternative, the net free cross-ventilation area may be reduced to $^1/_{300}$ when a vapor barrier having a transmission rate not exceeding 1 perm (5.7×10^{-11} kg/s · m^2 · Pa) is installed on the warm-in-winter side of the ceiling.

R806.3 Vent and insulation clearance. Where eave or cornice vents are installed, insulation shall not block the free flow of air. A minimum of a 1-inch (25 mm) space shall be provided between the insulation and the roof sheathing and at the location of the vent.

Q: A two-story house is being constructed. It has an $^8/_{12}$ pitched roof with an interior flat ceiling. The plan calls for R-61 roof vents above the level of the attic insulation. Does the code require soffit ventilation?

A: No. Soffit ventilation is optional. See Figure 8-9.

SOFFIT VENTS
FIGURE 8-9

Q: If soffit vents are not installed, does the attic insulation still need to be kept 1 inch away from the roof sheathing?

A: No.

Q: If soffit vents are installed, does the code require an air chute at each rafter or truss location to keep the insulation at least 1 inch away from the roof sheathing?

A: The code requires that when soffit vents are installed as a component of the total attic area ventilation, then the amount of air chutes is based on how much air could move past the air chute based on the design of the air chute. It may not be necessary to install an air chute at each truss space, but it will typically be installed this way due to the limited cost of time and material. See Figure 8-10.

CHUTE TO PROVIDE AIR SPACE FOR ATTIC VENTILATION
FIGURE 8-10

SECTION R807
ATTIC ACCESS

R807.1 Attic access. Buildings with combustible ceiling or roof construction shall have an attic access opening to attic areas that exceed 30 square feet (2.8 m^2) and have a vertical height of 30 inches (762 mm) or more.

The rough-framed opening shall not be less than 22 inches by 30 inches (559 mm by 762 mm) and shall be located in a hallway or other readily accessible location. A 30-inch (762 mm) minimum unobstructed headroom in the attic space shall be provided at some point above the access opening. See Section M1305.1.3 for access requirements where mechanical equipment is located in attics.

Q: Are there any conditions in the code that would permit an attic space without an access?

A: Yes, if the area or height does not exceed the limitations of Section R807.1.

NOTES

CHAPTER

ROOF ASSEMBLIES

9

SECTION R901 GENERAL

R901.1 Scope. The provisions of this chapter shall govern the design, materials, construction and quality of roof assemblies.

SECTION R902 ROOF CLASSIFICATION

R902.1 Roofing covering materials. Roofs shall be covered with materials as set forth in Sections R904 and R905. Class A, B or C roofing shall be installed in areas designated by law as requiring their use or when the edge of the roof is less than 3 feet (914 mm) from a property line. Classes A, B and C roofing required to be listed by this section shall be tested in accordance with UL 790 or ASTM E 108. Roof assemblies with coverings of brick, masonry, slate, clay or concrete roof tile, exposed concrete roof deck, ferrous or copper shingles or sheets, and metal sheets and shingles, shall be considered Class A roof coverings.

Q: Can you describe Class A, Class B and Class C fire ratings?

A: Class A roof coverings are effective against severe fire test exposures, Class B afford a moderate degree of fire protection and Class C coverings are effective against light fire test exposures when tested to UL 790, *Standard Test Method for Fire Tests of Roof Coverings*. The test covers the fire-resistance performance of roof coverings exposed to simulated fire sources originating from outside.

R903.4 Roof drainage. Unless roofs are sloped to drain over roof edges, roof drains shall be installed at each low point of the roof. Where required for roof drainage, scuppers shall be placed level with the roof surface in a wall or parapet. The scupper shall be located as determined by the roof slope and contributing roof area.

Q: Are gutters required by the code?

A: No.

SECTION R905 REQUIREMENTS FOR ROOF COVERINGS

R905.1 Roof covering application. Roof coverings shall be applied in accordance with the applicable provisions of this section and the manufacturer's installation instructions. Unless otherwise specified in this section, roof coverings shall be installed to resist the component and cladding loads specified in Table R301.2(2), adjusted for height and exposure in accordance with Table R301.2(3).

Q: If a roofing manufacturer's installation instructions are less restrictive than the building code, can we just follow the manufacturer's installation instructions. For example, the roofing manufacturer may require less extension of the ice protection up from the eave.

A: The roofing material needs to be installed according to requirements of both the code and the manufacturer's installation instructions according to Section R905.1.

R905.2 Asphalt shingles. The installation of asphalt shingles shall comply with the provisions of this section.

R905.2.1 Sheathing requirements. Asphalt shingles shall be fastened to solidly sheathed decks.

R905.2.2 Slope. Asphalt shingles shall be used only on roof slopes of two units vertical in 12 units horizontal (2:12) or greater. For roof slopes from two units vertical in 12 units horizontal (2:12) up to four units vertical in 12 units horizontal (4:12), double underlayment application is required in accordance with Section R905.2.7.

Q: I have two asphalt shingle packages, both from different manufacturers of shingles, and both say that the shingles can be installed over 1-inch nominal lumber wood decks. The code requires the asphalt shingles to be installed over solid sheathed decks. What now?

A: It is the intent of the code that asphalt shingles can be installed over solid wood boards. The important aspect is that the roofing material be provided with support at all areas, and the required nails are located in the wood boards, not in the spaces between the boards. Section R803 addresses wood structural sheathing and lumber used as solid sheathed decking. Refer to the Q & A for Section R803.

Q: Does the code allow a racking method for the application of roof shingles? See Figure 9-1.

A: Yes. The code does not specify if the shingles are to be laid continuously in a horizontal application or by a racking method as long as the shingles are still properly fastened. The manufacturer's installation instructions should be checked to see if there are any restrictions for this method of installation. The racking method will make it slightly more difficult to reduce the potential for color variations.

SHINGLES BEING APPLIED BY RACKING METHOD
FIGURE 9-1

R905.2.3 Underlayment. Unless otherwise noted, required underlayment shall conform to ASTM D 226 Type I, ASTM D 4869 Type I, or ASTM D 6757.

Self-adhering polymer modified bitumen sheet shall comply with ASTM D 1970.

Q: Is there any situation addressed in the code where asphalt shingles could be applied to exposed roof sheathing without an underlayment, such as based on square footage or use?

A: No. The code requires underlayment for all asphalt shingle installations. The underlayment provides increased moisture protection and provides a good walking surface during the installation of the shingles.

Q: If a manufacturer of asphalt shingles did not require an underlayment, would the code still require the underlayment?

A: Yes.

Q: I am designing a structure with a 2:12 sloped shed roof facing toward the north. I would like to cover all of the roof sheathing with the same self-adhering polymer modified membrane that was intended for the roof eave area prior to the installation of the asphalt shingles. Does the code allow for this practice?

A: Yes. The use of a self-adhering polymer barrier applied over all the roof sheathing is not addressed in the code. It is used for ice protection from the eaves' edges to a point 24 inches inside the exterior wall line. The installer should verify this application with the manufacturer of the shingles and of the self-adhering polymer prior to the installation. This practice is not uncommon, and is often used for commercial installations.

Q: Would the local building official need to approve this installation referred to in the previous question also?

A: No. The code requirement for the installation of underlayment is a minimum requirement, and if someone chose to use a product that was better than the minimum requirement, the code would not prevent it. The building official should be made aware of special concerns of the manufacturer of the shingles or of the underlayment material.

R905.2.5 Fasteners. Fasteners for asphalt shingles shall be galvanized steel, stainless steel, aluminum or copper roofing nails, minimum 12 gage [0.105 inch (3 mm)] shank with a minimum $^{3}/_{8}$-inch (10 mm) diameter head, ASTM F 1667, of a length to penetrate through the roofing materials and a minimum of $^{3}/_{4}$ inch (19 mm) into the roof sheathing. Where the roof sheathing is less than $^{3}/_{4}$ inch (19 mm) thick, the fasteners shall penetrate through the sheathing. Fasteners shall comply with ASTM F 1667.

Q: What information is contained in ASTM F 1667 that a roofer needs to know?

A: ASTM F 1667 is the *Standard Specification for Driven Fasteners: Nails, Spikes, and Staples.* Nails are identified by penny weight, length, head diameter and width, shank diameter (in inches) and approximate count per pound. They are further grouped based on type of material; coatings; barbed or smooth shank; point; and hand-driven versus mechanically driven.

Q: How are roofing nails identified in ASTM F 1667?

A: Roofing nails are identified as "Type I, Style 20."

Q: What is the minimum gauge of a roof nailing meeting that standard?

A: The minimum nail diameter noted is 0.106 inch, which is 12 gauge.

Q: How deep are the nails installed relative to the surface of the roof?

A: The nails should be flush with the surface of the shingle. If the nails are too high above the surface of the shingle, they may not allow the shingles to seal properly. If the nails are countersunk, they may not provide the holding power and may cut through.

Q: The section refers to ASTM F 1667. Does that standard contain any reference to staples for the application of asphalt shingles?

A: Yes. It identifies staples for roofing as Type IV staples. They are grouped by wire type, point, plating, coating, grouping (such as cohered), length, crown, gauge and ASTM identifier number.

Q: Does the code allow the use of staples for the installation of asphalt shingles?

A: The code specifically requires nails.

Q: If the shingle manufacturer allows the use of staples in the installation instructions, could the building official approve the use of staples for the shingles?

A: The building official could consider the installation using the provisions for alternate materials in Section R104.11.

Q: Are there any special requirements for pneumatic nailers?

A: No.

Q: Does the code address the condition of the roof decking prior to the installation of the underlayment or the shingles?

A: No. It should be applied to a dry deck. The manufacturer's installation instructions may have specific requirements related to the condition of the roof decking.

R905.2.6 Attachment. Asphalt shingles shall have the minimum number of fasteners required by the manufacturer. For normal application, asphalt shingles shall be secured to the roof with not less than four fasteners per strip shingle or two fasteners per individual shingle. Where the roof slope exceeds 20 units vertical in 12 units horizontal (167 percent slope), special methods of fastening are required. For roofs located where the basic wind speed per Figure R301.2(4) is 110 mph (49 m/s) or higher, special methods of fastening are required. Special fastening methods shall be tested in accordance with ASTM D 3161, Class F. Asphalt shingle wrappers shall bear a label indicating compliance with ASTM D 3161, Class F.

Q: It appears that most of the asphalt shingle manufacturers require the fasteners to be installed 5 $^{5}/_{8}$ inch up from the bottom of the shingle, which is slightly below the sealant strip. Can the fasteners be installed in the sealant strip?

A: No. Fastening into the sealant strip may not allow the shingles to be properly sealed down and it may contribute to blowoffs.

Q: Can the fasteners be installed above the sealant strip?

A: No. The fasteners are installed below the sealant strip to pull the shingle surface into the sealant.

Q: Does the code require metal drip edge to be installed with an asphalt shingle installation?

A: In Section R905.1 it specifically requires the roof covering material to be installed in accordance with the code and the manufacturer's installation instructions. Many shingle manufacturers require drip edge as part of their installation instructions. For dwelling constructed under the IBC, such as attached Group R-3 dwellings (six units and greater back-to-back) and Group R-2 apartments, it is required in Section 1507.2.9.3.

Q: If the drip edge is required by the manufacturer, is it installed over the underlayment or under the underlayment?

A: Refer to the shingle manufacturer's installation instructions. The drip edge is typically installed over the underlayment at the rake, and under the underlayment (or ice protection) at the eave.

R905.2.7.1 Ice barrier. In areas where there has been a history of ice forming along the eaves causing a backup of water as designated in Table R301.2(1), an ice barrier that consists of a least two layers of underlayment cemented together or of a self-adhering polymer modified bitumen sheet, shall be used in lieu of normal underlayment and extend from the lowest edges of all roof surfaces to a point at least 24 inches (610 mm) inside the exterior wall line of the building.

> **Exception:** Detached accessory structures that contain no conditioned floor area.

Q: Why does the code have an exception for the ice barrier in detached accessory structures that contain no conditioned space?

A: The ice protection is required for dwellings located in cold weather climates where the heat loss through the roof sheathing creates a situation where the snow melts down the roof and refreezes at the eaves, creating ice dams. The further melting of the snow can then back up under the shingles. The nonconditioned (nonheated) space of the detached garage is exempt from this requirement.

Q: Would the ice protection be required if the shingle manufacturer required this ice protection at the eaves?

A: Yes. The roof installation needs to meet the requirements of the shingle manufacturer per Section R905.1.

Q: What is considered conditioned floor area?

A: Conditioned floor area is a space within a building that is provided with heating and/or cooling equipment capable of maintaining 50°F during the heating season.

Q: If a detached garage with no conditioned space had a shingled roof installed without the ice protection, would the shingles be in compliance with the code if a heat source, such as a space heater, was installed at a later date?

A: If a heating source was installed, the shingles would no longer be in compliance with the code. The owner may want to install insulation and ventilation in the attic space to help prevent any heat build-up in the attic space that may contribute to the snow melting.

Q: Our staff believes that the ice barrier requirement for the protection to extend from the lowest edge of the eave to a point 24 inches inside the exterior wall line is based on an average roof overhang of 24 inches to 36 inches. If a roof projects 6 or 8 feet over an unconditioned open porch, such as in Figure 9-2, does the ice barrier still need to extend 24 inches inside the exterior wall line of the structure?

A: Yes. The provisions for ice protection do not differentiate between different roof slopes or amount of overhang.

R905.2.8.2 Valleys. Valley linings shall be installed in accordance with the manufacturer's installation instructions before applying shingles. Valley linings of the following types shall be permitted:

1. For open valley (valley lining exposed) lined with metal, the valley lining shall be at least 24 inches (610 mm) wide and of any of the corrosion-resistant metals in Table R905.2.8.2.
2. For open valleys, valley lining of two plies of mineral surfaced roll roofing, complying with ASTM D 3909

or ASTM D 6380 Class M, shall be permitted. The bottom layer shall be 18 inches (457mm) and the top layer a minimum of 36 inches (914 mm) wide.

3. For closed valleys (valley covered with shingles), valley lining of one ply of smooth roll roofing complying with ASTM D 6380 Class S Type III, Class M Type II, or ASTM D 3909 and at least 36 inches wide (914 mm) or valley lining as described in Items 1 and 2 above shall be permitted. Specialty underlayment complying with ASTM D 1970 may be used in lieu of the lining material.

ICE BARRIER PROTECTION ABOVE FRONT ENTRY
FIGURE 9-2

Q: Could a self-adhering modified bitumen be used as the underlayment in the valley?

A: It would need to meet the ASTM standard specified for the application in Section R905.2.8.2. Most modified bitumen materials cannot remain exposed to sunlight.

R905.2.8.4 Sidewall flashing. Flashing against a vertical sidewall shall be by the step-flashing method.

R905.2.8.5 Other flashing. Flashing against a vertical front wall, as well as soil stack, vent pipe and chimney flashing, shall be applied according to the asphalt shingle manufacturer's printed instructions.

Q: Does the code specify how high above the roof sheathing the sidewall flashing needs to extend?

A: No, although it may be noted in the shingle manufacturer's installation instructions. See Figure 9-3.

SIDEWALL FLASHING
FIGURE 9-3

Q: Could a continuous section of flashing properly sealed be used in place of step flashing?

A: No. Step flashing is installed in between each course as the shingle is installed up the sidewall, allowing the rainwater to run over the top of each shingle and down the roof.

SECTION R907
REROOFING

R907.1 General. Materials and methods of application used for re-covering or replacing an existing roof covering shall comply with the requirements of Chapter 9.

Exception: Reroofing shall not be required to meet the minimum design slope requirement of one-quarter unit vertical in 12 units horizontal (2-percent slope) in Section R905 for roofs that provide positive roof drainage.

R907.2 Structural and construction loads. The structural roof components shall be capable of supporting the roof covering system and the material and equipment loads that will be encountered during installation of the roof covering system.

R907.3 Re-covering versus replacement. New roof coverings shall not be installed without first removing existing roof coverings where any of the following conditions occur:

1. Where the existing roof or roof covering is water-soaked or has deteriorated to the point that the ex-

isting roof or roof covering is not adequate as a base for additional roofing.

2. Where the existing roof covering is wood shake, slate, clay, cement or asbestos-cement tile.
3. Where the existing roof has two or more applications of any type of roof covering.
4. For asphalt shingles, when the building is located in an area subject to moderate or severe hail exposure according to Figure R903.5.

Exceptions:

1. Complete and separate roofing systems, such as standing-seam metal roof systems, that are designed to transmit the roof loads directly to the building's structural system and that do not rely on existing roofs and roof coverings for support, shall not require the removal of existing roof coverings.
2. Installation of metal panel, metal shingle, and concrete and clay tile roof coverings over existing wood shake roofs shall be permitted when the application is in accordance with Section R907.4.
3. The application of new protective coating over existing spray polyurethane foam roofing systems shall be permitted without tear-off of existing roof coverings.

R907.4 Roof recovering. Where the application of a new roof covering over wood shingle or shake roofs creates a combustible concealed space, the entire existing surface shall be covered with gypsum board, mineral fiber, glass fiber or other approved materials securely fastened in place.

Q: Can a third layer of asphalt shingles be installed over two existing layers of asphalt shingles?

A: No. If the roof already has two layers of asphalt shingles, those would need to be removed prior to the installation of the new shingles.

Q: Could one layer of shingles be removed from the existing two layers of asphalt shingles so one new layer could be applied over it?

A: If the building is located in an area subject to moderate or severe hail exposure according to Figure R903.5 in the code, then only one layer is permitted. Also, the installer should check with the manufacturer of the shingles about this installation. Besides being somewhat of a difficult removal process, which may end up destroying the initial layer of shingles, it may not meet the requirements of the manufacturer.

Q: On a reroofing project, we removed the existing shingles down to the felt paper and the ice protection. The felt paper came up very easy. The self-adhering ice protection barrier could not be removed from the surface of the roof sheathing. Can the ice protection remain in place?

A: The self-adhering polymer modified ice protection is not made to be removed once applied and allowed to seal. Although this type of product has only been used for a decade or so, a roofing contractor may run into this situation on newer roofs that have been damaged during a severe hail storm, for instance. In many cases, if this product is forcibly removed, it will also damage the roof sheathing. It would be best to refer to the manufacturer for recommendations, followed by a visit to the local building official to see how this could be resolved.

Q: If the ice protection is not removed during a reroofing application, does it now leave a lot of extra holes in the membrane after the old shingles are removed?

A: Relatively speaking, the area left by the nails that are removed are very minor, and often the ice protection will reseal the nail opening. This area will also be once again covered from direct contact with the rain.

NOTES

CHIMNEYS AND FIREPLACES

CHAPTER

10

SECTION R1001 MASONRY FIREPLACES

R1001.1 General. Masonry fireplaces shall be constructed in accordance with this section and the applicable provisions of Chapters 3 and 4.

R1001.2 Footings and foundations. Footings for masonry fireplaces and their chimneys shall be constructed of concrete or solid masonry at least 12 inches (305 mm) thick and shall extend at least 6 inches (152 mm) beyond the face of the fireplace or foundation wall on all sides. Footings shall be founded on natural, undisturbed earth or engineered fill below frost depth. In areas not subjected to freezing, footings shall be at least 12 inches (305 mm) below finished grade.

Q: A plan for a residential dwelling indicates a masonry fireplace with a masonry chimney at the exterior wall of the dwelling. The code calls for a 12-inch-deep footing with 6-inch projection for this masonry fireplace and chimney. The exterior wall footings are indicated on the plan as 20 inch wide and 8 inches deep. Can the footing for the masonry fireplace and chimney be poured continuous with the exterior wall footing, or does it need to be poured independent of it?

A: The footings can be poured continuous, although there will be a 4-inch step in the footing where the transition occurs between the two footings. The footings should be placed on undisturbed earth or engineered fill.

R1001.9 Hearth and hearth extension. Masonry fireplace hearths and hearth extensions shall be constructed of concrete or masonry, supported by noncombustible materials, and reinforced to carry their own weight and all imposed loads. No combustible material shall remain against the underside of hearths and hearth extensions after construction.

Q: Most builders in our area form up the fireplace hearth extension with wood framing, reinforce as needed and then pour the concrete. Often during the final inspection we check the underside of the hearth extension from the basement level looking up, and if the formwork is still in place we ask for it to be removed. Is this something that should be done at the final inspection?

A: The code requires that no combustible material remain against the underside of the hearth extension. The self-supporting capabilities of the hearth are determined by the design and proper installation of the hearth materials. This should be verified at the final inspection. See Figure 10-1.

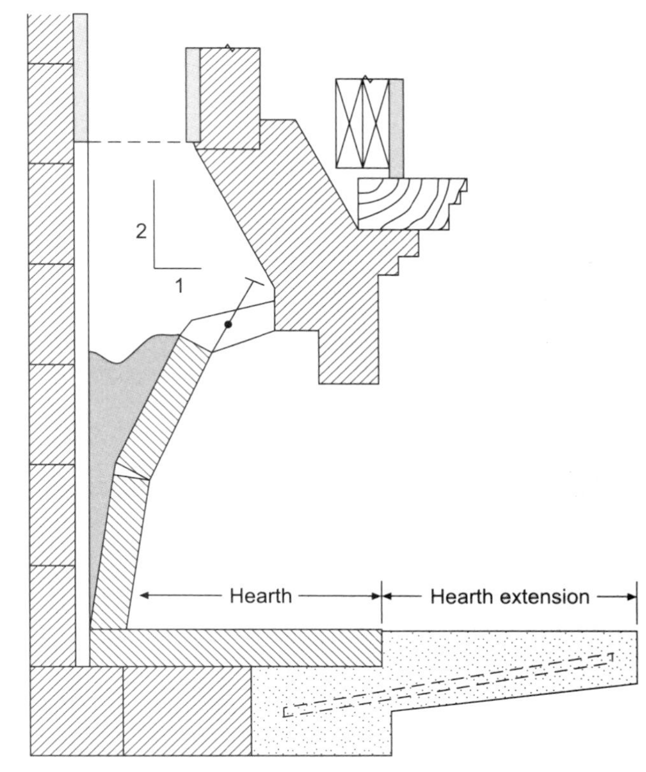

FIREPLACE HEARTH EXTENSION
FIGURE 10-1

Q: Once the formwork is removed from the bottom of the hearth extension, we check for clearance between the concrete hearth extension and the wood floor joists on the three sides. What is the minimum clearance between the joists and the hearth extension?

A: The code does not specifically require clearance between the sides of the concrete hearth extension and the wood floor joists.

R1001.11 Fireplace clearance. All wood beams, joists, studs and other combustible material shall have a clearance of not less than 2 inches (51 mm) from the front faces and sides of masonry fireplaces and not less than 4 inches (102 mm) from the back faces of masonry fireplaces. The air space shall not be filled, except to provide fire blocking in accordance with Section R1001.12.

4. Exposed combustible mantels or trim may be placed directly on the masonry fireplace front surrounding the fireplace opening providing such combustible materials are not placed within 6 inches (152 mm) of a fireplace opening. Combustible material within 12 inches (306 mm) of the fireplace opening shall not project more than $^1/_8$ inch (3 mm) for each 1-inch (25 mm) distance from such an opening.

Q: If a piece of combustible wood trim is installed on the front side of the masonry fireplace 6 inches away from the fireplace opening, how far out could this piece of combustible wood trim project?

A: It can project $^1/_8$ inch for each 1 inch distance from the opening. In other words, $^1/_8$ inch times 6 inches equals a ¾-inch projection. See Figure 10-2.

Q: When a masonry chimney for a masonry fireplace passes through a roof on the interior of a house, what is the minimum clearance to combustible material and how can this be constructed?

A: In Section R1003.18, "Chimney clearances," the code requires a minimum air space clearance to combustibles of 2 inches. The roof framing will need to maintain the 2-inch clearance from the masonry chimney. Any connection between the wood framing and the masonry chimney would need to be by the use of metal or some other noncombustible material.

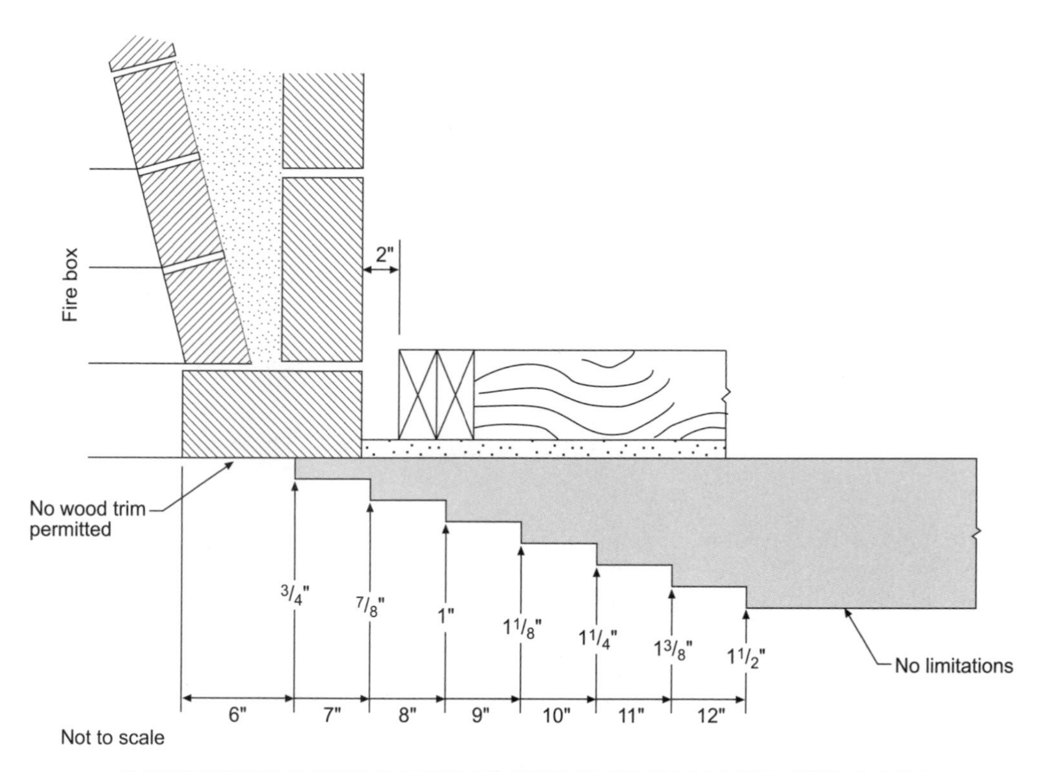

COMBUSTIBLE WOOD TRIM AT FRONT OF MASONRY FIREPLACE
FIGURE 10-2

SECTION R1004 FACTORY-BUILT FIREPLACES

R1004.1 General. Factory-built fireplaces shall be listed and labeled and shall be installed in accordance with the conditions of the listing. Factory-built fireplaces shall be tested in accordance with UL 127.

R1004.2 Hearth extensions. Hearth extensions of approved factory-built fireplaces shall be installed in accordance with the listing of the fireplace. The hearth extension shall be readily distinguishable from the surrounding floor area.

Q: Is a 16-inch hearth extension required for a wood-burning metal fireplace?

A: A hearth extension for a factory-built listed and labeled fireplace needs to be installed according to the listing.

Q: Could a listed factory-built fireplace use a masonry chimney?

A: Yes, unless specifically prohibited by the listing for the fireplace.

Q: If the masonry chimney was approved by the listing, what provisions in the code apply to the chimney itself?

A: Basically, it is the same as for a masonry fireplace (footing sizes, clearances, etc.), and the chimney provisions of Chapter 18 of the IRC.

NOTES

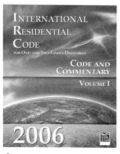

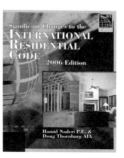